Advances in Experimental Medicine and Biology

Volume 822

More information about this series at http://www.springer.com/series/5584

Panayiotis Vlamos • Athanasios Alexiou
Editors

GeNeDis 2014

Neurodegeneration

Springer

Editors
Panayiotis Vlamos
Department of Informatics
Ionian University
Corfu, Greece

Athanasios Alexiou
Department of Informatics
Ionian University
Corfu, Greece

ISSN 0065-2598 ISSN 2214-8019 (electronic)
ISBN 978-3-319-08926-3 ISBN 978-3-319-08927-0 (eBook)
DOI 10.1007/978-3-319-08927-0
Springer Cham Heidelberg New York Dordrecht London

Library of Congress Control Number: 2014954991

Printed on acid-free paper

Springer is part of Springer Science+Business Media (www.springer.com)

Contents

GeNeDis 2014 Information: Springer AEMB

GeNeDis 2014 Overview

The 1st World Congress on Geriatrics and Neurodegenerative Disease Research, GeNeDis 2014 focused on recent advances in Geriatrics and Neurodegeneration, ranging from basic science to clinical and pharmaceutical developments, providing also an international forum for the latest scientific discoveries, medical practices, and care initiatives. Leading scientists and experts, students, physicians and nurses, professionals as well as industries representatives, and many other participants discussed the latest major challenges, new drug targets, the development of new biomarkers, imaging techniques, novel protocols for early diagnosis of neurodegenerative diseases, bioinformatics methods, and several other scientific achievements. While the European Union aims to strengthen the research and innovation towards Horizon 2020, the translational research for healthy aging and neurodegeneration will remove any barrier on multidisciplinary collaboration. Novel approaches including biomarkers, stem cell therapy, protein misfolding, immunotherapy, as well as developments in our understanding of the genetics, the molecular mechanisms and signaling pathways contributing to neuronal dysfunction, nanotechnological products, and innovative computational methods will offer new research directions and strategies on future preclinical and clinical studies on neurodegenerative diseases and further improvement of quality services on rehabilitation and health education. Advanced information technologies have been discussed concerning the various research, implementation, and policy, as well as European and global issues in the funding of long-term care and medico-social policies regarding elderly people.

We can resume the major objectives of the 1st World Congress on Geriatrics and Neurodegenerative Disease Research, GeNeDis 2014 as follows:

- To promote the scientific and research results to the research community through sharing and exchanging ideas, experiences, and expectations focusing on Health Aging and Mental Wellness in the new digital era.

- To explore the beneficial (and not only) impact of scientific and technological achievements and future challenges that may affect the Health Aging and Mental Wellness process.
- To enhance the co-operation and the exchange of experiences and resources among organizers of events and the research community, increasing further the European dimension and added value of the activities of the Europe 2020 and HORIZON 2020.
- To increase the interest of all the stakeholders in contributing to the research in the field, while mental health is a basic human right, and is fundamental to all human and social progress. It is a prerequisite to a happy and fulfilled life for individual citizens, starting at birth, for functioning families and for societal cohesion.

The Organizers

- Ionian University, Department of Informatics, Bioinformatics and Human Electrophysiology, Greece
- Ionian University, Department of Informatics, Computational Modeling Laboratory, Greece
- Science View, Member of the European Science Journalists EUSJA, Greece

The Co-organizers

- Hellenic Open University, School of Science and Technology, Greece
- University of Salento, Department of Innovation Engineering, Laboratory of Applied Mechanics, Section of Mechatronics, Italy
- Ionian University, Department of History, M.Sc., in Historical Research, Teaching and New Technologies, Greece
- Region of Ionian Islands, Greece

Scientific Program Committee

Conference Chair

Vlamos Panayiotis, Department of Informatics, Ionian University, Greece

Conference Vice-Chair

Alexiou Athanasios, Department of Informatics, Ionian University, Greece

- Acosta Sandra, Institut de Recherche Interdisciplinaire en Biologie Humaine et Moléculaire, Belgium
- Aigner Ludwig, Paracelsus Medizinische Privatuniversität, Austria

- Alexopoulos Dimitrios, Dental Surgeon, Editorial Director Dentorama, Dental Journal and Dental Tribune, Hygeia Hospital
- Anagnostopoulos Dimosthenis, Harokopeio University of Athens, Greece
- Ashraf Md Ghulam, King Fahd Medical Research Center, King Abdulaziz University, Saudi Arabia
- Bilensoy Erem, Hacettepe Üniversitesi Eczacılık Fakültesi Farmasötik Teknoloji Anabilim Dalı, Turkey
- Bottalico Barbara, University of Pavia, European Center for Law, Science and New Technologies, University of Pavia, Italy
- Chrissikopoulos Vassileios, Ionian University, Greece
- Chatzidaki Georgia, Corfu General Hospital, Greece
- De Castro Fernardo, Grupo de Neurobiología del Desarrollo-GNDe Hospital Nacional de Parapléjicos-SESCAM, Spain
- Delabar JM, Université Paris Diderot, France
- Demetzos Costas, Dept. of Pharmaceutical Technology, Faculty of Pharmacy, National and Kapodistrian University of Athens, Greece
- D'Mello Santosh R., University of Texas, USA
- Exarchos Themis, Department of Computer Science, University of Ioannina, Greece
- Fjeldsted Katrin, President Standing Committee of European Doctors
- Fotiadis Dimitrios, University of Ioannina, Greece
- Geoghegan-Quinn Maire, Commissioner of Research, Innovation and Science, EU
- Georgiadis Adonis, Minister For Health, Greece
- Giannakopoulos Panteleimon, Professor, Chairman of the Department of Mental Health and Psychiatry, University Hospitals of Geneva, Switzerland
- Ghiatas Abraham, University of Texas, USA
- Gkigkitzis Ioannis, Departments of Mathematics and Biomedical Physics, East Carolina, USA
- Gonos Efstathios, National Hellenic Research Foundation, Institute of Biological Research and Biotechnology, Greece
- Hadjinicolaou Maria, Hellenic Open University, Greece
- Haranas Ioannis, VDG Research and Development in Electromagnetic Theory Applications, Department of Mathematics, East Carolina University, USA
- Huotilainen Minna, University of Helsinki, Finland
- Ifantis Ilias, Hellenic Naval Academy, Piraeus, Greece
- Karabatsas Kostantinos, Governor of General Hospital Achillopouleion of Volos
- Karalis Vangelis, Faculty of Pharmacy, National Kapodistrian University of Athens, Greece
- Kariotou Foteini, Hellenic Open University, Greece
- Kontogeorgis Georgios, Technical University of Denmark
- Koumoutsakos Georgios, Commissioner, EU
- Knyazeva Maria, Lausanne University Hospital, Switzerland
- Lefkimmiatis Konstantinos, Department of Physiology Anatomy and Genetics, University of Oxford

- Logothetis Nikos, Max-Planck-Institute for Biological Cybernetics, Germany
- Luhan Petru Constantin, Commissioner, EU
- Maramabaud Philippe, The Feinstein Institute for Medical Research
- Matsas Rebecca, Hellenic Pasteur Institute, Greece
- Näätänen Risto, University of Helsinki, Finland
- Oertel Wolfgang, Neurological Clinic Institution, Philipps-Universitat Marburg, Germany
- Orban Guy, KU Leuven, Belgium
- Pasinetti Giulio, Icahn School of Medicine at Mount Sinai, Department of Neurology and Friedman Brain Institute, New York
- Pateli Adamantia, Ionian University, Greece
- Priyadarshini Madhusmita, University of Helsinki, Finland
- Protopappas Vasilios, University of Ioannina, Greece
- Psiha Maria, Ionian University, Greece
- Ratana Rajiv R., Burke Medical Research Institute, Cornell University, USA
- Robakis Nikolaos, Mount Sinai Hospital, New York
- Sakka Paraskevi, Athens Alzheimer's Association, Dementia Society, Greece
- Serrano Alberto, University Autonoma Madrid, Spain
- Sattler Wolfgang, University of Graz, Austria
- Simou Panagiota, Department of Audio and Visual Arts, Ionian University, Greece
- Sioutas Spyros, Ionian University, Greece
- Tabrez Shams, King Fahd Medical Research Center, King Abdulaziz University, Saudi Arabia
- Takashima Akihito, Department of Aging Neurobiology
- Tarnanas Ioannis, Universität Bern, Germany
- Tiligadis Konstantinos, Department of Audio and Visual Arts, Ionian University, Greece
- Tripoliti Evi, University of Ioannina, Greece
- Tsamis Konstantinos, University of Ioannina, Greece
- Tsantili-Kakoulidou Anna, President of the Society of Hellenic Medicinal Chemistry, Greece
- Tsolaki Magda, Aristotle University of Thessaloniki, Greece
- Tsoumakos Dimitrios, Ionian University, Greece
- Velayudhan Latha, Department of Health Science, University of Leicester, England
- Van der Jeugd Anneke, Laboratory of Biological Psychology, KU Leuven, Belgium
- Villringer Arno, Max Planck Institute for Human Cognitive and Brain Sciences, Germany
- Volikas Kimon, University of Nicosia, Cyprus
- Wolozin Benjamin, Department of Pharmacology Boston University Alzheimer's Disease Center
- Xiang Zhao, Neuroscience Center, University of Helsinki, Finland
- Xing Wei, Scientific Computing/PI, University of Manchester, United Kingdom

- Yapijakis Christos, Department of Oral and Maxillofacial Surgery, University of Athens Medical School "Attikon" Hospital, Greece
- Zaidi Syed Kashif, Center of Excellence in Genomic Medicine Research, King Abdulaziz University, Saudi Arabia
- Zaganas Ioannis, Department of Neurology, Medical School, University of Crete, Greece
- Zouganelis George, Bournemouth and Poole College, United Kingdom

Keynote Lectures

- Professor Nikolaos Robakis, Professor of Psychiatry, Neuroscience and experimental therapeutics. First A.P. Slaner Professor for Alzheimer disease research. Mount Sinai School of Medicine, NYU, USA, Title: Molecular neuropathology of Alzheimer dementia and therapeutic approaches
- Professor George Paxinos, A.O., D.Sc., F.A.S.S.A., F.A.A., N.H.M.R.C., Australia Fellow, Neuroscience Research Australia, Barker St and Hospital Rd, Randwick, Sydney, Australia, Title: Brain, Behaviour and Evolution
- Professor Vasilis Ntziachristos, Chair Institute for biological and medical imaging (IBMI) Technische Universität München, Germany, Title: Biomedical optics and optoacoustics as a model for scientific discovery and economic growth
- Dr. Akihito Takashima, Department of Aging Neurobiology, National Center for Geriatrics and Gerontology, Japan, Title: Toxic tau aggregation in AD
- Professor Benjamin Wolozin, Department of Pharmacology Boston University Alzheimer's Disease Center, USA, Title: Stress granules and Neurodegeneration: A new paradigm for mechanisms of neurodegeneration
- Professor Panteleimon Giannakopoulos, Chairman of the Department of Mental Health and Psychiatry, University Hospitals of Geneva, Switzerland, Title: Mental health in a global world: moving from illness to individual vulnerability
- Dr. Efstathios Gonos, Director of Research IUBMB, National Hellenic Research Foundation, Institute of Biology, Medicinal Chemistry and Biotechnology, Greece, Title: Proteasome activation delays ageing and minimizes deficiencies underlying neurodegenerative diseases
- Dr. Ioannis Haranas, VDG Research and Development in Electromagnetic Theory Applications, Department of Mathematics, East Carolina University, USA, Title: Respiratory Particle Deposition Probability due to Sedimentation with Variable Gravity and Electrostatic Forces
- Professor Dimitrios Fotiadis, Unit of Medical Technology and Intelligent Information Systems, University of Ioannina, Title: Sequence patterns mediating functions of disordered proteins
- Professor Arno Villringer, Max Planck Institute for Human Cognitive and Brain Sciences, Germany, Title: The path from obesity and hypertension to dementia
- Professor Constantin Bouras, University Hospitals, Geneva, Department of Psychiatry, Switzerland, Title: Individual severity of AD-type lesions, Aß oligomers and comorbidity in the brain aging

Invited Talks: Workshops and Round Tables

- Professor Theodosia Vallianatou, Department of Pharmaceutical Chemistry, School of Pharmacy, University of Athens, Greece, Title: The impact of physicochemical and molecular properties in drug design. Navigation in the "druglike" chemical space
- Professor Costas Demetzos, Faculty of Pharmacy, National and Kapodistrian University of Athens, Title: Advanced Drug Delivery nano Systems: perspectives and regulatory issues
- Dr. Paraskevi Sakka, Neurologist-Psychiatrist, Director at Hygeia Hospital, Neurodegenerative Brain Conditions – Memory Clinic, Chairwoman at Athens Association of Alzheimer's Disease, and Related Disorders, Greece, Title: Non-pharmacological treatments of dementia. The encouraging results of SOCIABLE

Chapter 1
Molecular Neuropathology of Alzheimer Dementia and Therapeutic Approaches

Nikolaos K. Robakis

Alzheimer's disease (AD) is a neurodegenerative disorder of the central nervous system characterized by progressive loss of neurons in limbic and association cortices affecting almost all cognitive functions and memory. Additional neuropathological hallmarks of AD brains include large numbers of amyloid plaques (APs) containing fibers of Aβ peptides and neurofibrillary tangles (NFTs) consisting of overphosphorylated tau protein. Sporadic AD (SAD) is the most common form of dementia and accounts for more than 90 % of all cases. The etiology of SAD is complex and may be driven by both environmental and genetic factors. Presence of ApoE4 allele is the most important genetic risk factor for SAD although in the last 5 years additional genes were identified with smaller effects on the disorder. Several theories have been developed to explain the molecular basis of AD but none of these seems to offer a satisfactory explanation for the neurodegenerative phenotype of the disorder.

Familial AD (FAD) represents about 5 % of all AD cases. In contrast to sporadic AD, FAD is caused by specific mutations in several genes including those of the amyloid precursor protein (APP), presenilin1 (PS1) and presenilin2 (PS2). Currently over 200 FAD mutations have been mapped to PSs and about 20 to APP. PS proteins are important components of the γ-secretase system that processes cell surface proteins and receptors including APP. Importantly, γ-secretase processing of APP is involved in the production of Aβ peptides that aggregate to form the amyloid fibers used to define the disorder. In addition, γ-secretase stimulates production of peptides that function in cell signaling and gene expression, key events in neuronal survival. Thus, by inhibiting the activity of γ-secretase and production of signaling peptides FAD mutations may interfere with the flow of cellular information. In this keynote talk we will discuss mechanisms by which genetic mutations lead to neurodegeneration and dementia and will examine genetic and environmental factors involved in SAD. In addition, we will illuminate recent efforts for the development of second-generation diagnostics and therapeutics.

N.K. Robakis (✉)
Department of Psychiatry, Neuroscience and Experimental Therapeutics,
Mount Sinai School of Medicine, NYU, New York, NY, USA
e-mail: nikos.robakis@mssm.edu

P. Vlamos, A. Alexiou (eds.), *GeNeDis 2014*, Advances in Experimental Medicine and Biology 822, DOI 10.1007/978-3-319-08927-0_1

Chapter 2
Toxic Tau Aggregation in AD

Akihiko Takashima

Keywords Granular tau oligomer • Tau fibril • Neuron loss • Clinical progression

The pathological hallmarks of Alzheimer's disease (AD) are extracellular β-amyloid deposition and intracellular tau inclusions. While β-amyloid deposition does not correlate with clinical progression of AD, the diffusion of neurofibrillary tangles (NFTs) from the entorhinal cortex to the neocortex, followed by neuronal and synapse loss, matches well with the clinical progression of AD symptomatology. Therefore, blocking the formation of NFTs is considered to be a promising approach to halt the progression of AD dementia.

Our analysis of the temporal formation of Tau fibrils in vitro showed that there are different and distinct forms of tau aggregates (soluble tau oligomer, and granular tau oligomer) that precede Tau fibril formation.

Analysis of our P301L-Tau Tg mouse model suggested that toxicity of Tau aggregates could be attributed to granular tau. To test this further, we attempted to reduce formation of granular tau oligomer by screening the chemical compound X1 that was shown to inhibit the formation of granular tau. Interestingly, oral administration of X1 to P301L-Tau mice resulted in reduced neuronal loss, accompanied by a reduction of the levels of sarkosyl-insoluble tau, as compared to control vehicle treatment. Together, these studies offer novel insights into Tau aggregation pathology; they strongly suggest that granular tau oligomers represent a toxic tau aggregate and that X1 may be a promising compound for blocking AD progression.

2.1 Introduction

In Alzheimer's disease (AD), β-amyloid deposition and neurofibrillary tangles (NFTs), composed of hyperphosphorylated tau fibrils, are considered to be a major pathological feature. Evidence from the genetic studies of familial AD has given rise to the β-amyloid hypothesis, in which β-amyloid is proposed as the cause

A. Takashima (✉)
Department of Aging Neurobiology, National Center for Geriatrics and Gerontology, 35 Morioka, Oobu-shi, Aichi, Japan
e-mail: kenneth@ncgg.go.jp

P. Vlamos, A. Alexiou (eds.), *GeNeDis 2014*, Advances in Experimental Medicine and Biology 822, DOI 10.1007/978-3-319-08927-0_2

of AD; thus, the reduction of β-amyloid has been viewed as a potential therapeutic target for AD [1]. However, most studies targeting β-amyloid have failed in phase III clinical trial, as discussed extensively elsewhere [2]. A major explanation for such failures may be that such therapies have been tested in patients with early- to mid-stage AD when disease progression is already relatively advanced. In fact, when first AD symptoms are reported or detected, it seems that damage is irreversible and not amenable to slowing down or blockade. Accordingly, researchers are attempting to identify very early stages of disease, before conversion to, that may be therapeutically targetable; that is, they are now seeking an early "window of therapeutic opportunity." Even so, this approach depends on differentiating between the trigger and bullet in AD, and their relationship to β-amyloid. At the same time, researchers are becoming aware that alternative target for halting clinical progression even in early to moderate AD may be necessary.

2.2 Cause of Clinical Progression in AD

Tacitly, β-amyloid deposition may be the consequence, rather than the cause of disease. Indeed, all causative genes in familial AD relate to β-amyloid production, and the A673T protective APP mutation, which reduces Ab production [3]; A673T carriers show better cognitive scores than noncarriers. However, these findings suggest that not only β-amyloid is a cause of cognitive dysfunction, but also that dysfunction of APP or the products of other causatively linked genes are likely to be involved in β-amyloid production and cognitive dysfunction. Although β-amyloid toxicity has been observed in cultured neurons, mutant APP overexpressing mice do not show neuronal loss, but rather show memory impairment and senile plaques. Interestingly, immunotherapy against β-amyloid reduces β-amyloid deposition, and improves memory dysfunction in mutant APP overexpressing mouse model; however, immunotherapy does not halt clinical progression in AD patients, although immunotherapy reduces β-amyloid deposition in AD patients. Taken together, these observations suggest that β-amyloid is involved in Ab deposition and neuronal dysfunction, but not directly in neuronal death in the mature brain, explaining why β-amyloid targeting therapy does not prevent clinical progression of AD. It would appear that neuronal death may be a critical factor in clinical progression of the disease, indicating the need for alternative therapeutic targets and approaches.

2.3 Relationship Between NFTs and Clinical Progression of AD

NFTs are the other pathological hallmark of AD. While β-amyloid deposition is only seen in AD, NFTs are associated with several neurodegenerative diseases, the so-called tauopathies. Because all tauopathies are accompanied by NFTs and neuronal dysfunction, NFTs are considered to be a common pathological marker for a range of neurological disorders. In these diseases, the rate of neuronal loss exceeds the occurrence of NFTs, suggesting that NFT formation and neuron death share a common underlying mechanism [4, 5]. This hypothesis is strongly supported by the discovery of a tau gene mutation in patients with frontotemporal dementia with parkinsonism linked to chromosome 17 (FTDP-17) [6–8]. The FTDP-17-associated tau gene mutation is the causal factor in FTDP-17, a dementing disease characterized by NFT formation and neuron loss. Analyses of mutated tau in FTDP-17 conclusively demonstrated that tau dysfunction or abnormality alone induces the neurodegeneration characterized by NFTs and neuronal death that ultimately leads to clinical dementia.

2.4 NFTs and Neuronal Dysfunction

Braak and colleagues defined the progression of AD in terms of six stages, based on the distribution of NFTs in the brain [9]. In Braak stage I, NFTs are observed in the transentorhinal cortex and the CA1 region of the hippocampus. The number of NFTs increases in Braak stage II, and Braak stages I and II together are called the transentorhinal stage. Brains from normal non-demented aged subjects are also often categorized as Braak stages I and II. In Braak stages III and IV, called the limbic stage, many ghost tangles appear in the entorhinal cortex, and NFTs are found throughout the entire limbic system, including hippocampal regions CA1-4 and the amygdala. In the limbic stage, patients show various AD-specific symptoms, such as memory impairment, reduced spatial cognition, and reduced motivation, ascribable to neural dysfunction in the limbic system. In Braak stages V and VI, called the isocortical stage, NFTs are present in the cerebral cortex, where they impair neural function and cause dementia. The increasing spread of NFTs from the transentorhinal cortex to the limbic system and finally to the cerebral cortex correlates with increasing impairment of brain function. Samuel and colleagues reported that the number of NFTs in CA1, subiculum, and CA4 of the hippocampal formation correlates with the degree of dementia [10]. Therefore, the distribution of NFTs is now considered to correlate better with disease progression in AD than β-amyloid production and deposition.

2.5 Intermediate Form of NFT

Mice that overexpress P301L mutant tau under the regulation of a tetracycline-inducible promoter display age-related NFTs, neuronal death, and behavioral deficits. Although inhibition of mutant tau in these mice blocks neuronal death and improves memory, NFTs continue to form [11–13]. This suggests that NFTs are not themselves toxic, but rather, that the mechanism of NFT formation is shared by the process underlying neuronal death and neuronal dysfunction.

To understand a process of NFT formation that links with neuronal death, we first need to know how monomeric tau forms fibrillar tau. Anionic surfactants accelerate fibril formation of tau protein in vitro, and fibril formation can then be monitored by thioflavin (ThT) fluorescence, which recognizes protein aggregations with a β-sheet conformation. To track structural changes in tau in solution, and to understand the relationship between different tau aggregates, Maeda and colleagues investigated how tau assembly in vitro changes over time by measuring ThT fluorescence, using atomic force microscopy (AFM) [14].

Before ThT fluorescence can be detected, tau forms oligomers (dimers to octamers) with increasing time of incubation [15]. Although these tau oligomers could not be observed under AFM, they were detected by SDS-PAGE performed under nonreducing conditions. Under reducing conditions, however, tau oligomers, consisting of three or more tau molecules, could no longer be detected, suggesting that tau oligomers form through disulfide bonds and through other SDS-resistant tau–tau associations. As ThT fluorescence increases, two forms of tau aggregates can be observed with AFM: a granular tau oligomer and a fibrillar tau aggregate. Laser light-scattering analysis of sucrose-gradient purified granular tau oligomers indicated that a single granular tau oligomer consists of about 40 tau molecules. Since higher concentrations of granular tau oligomer induce the formation of tau fibrils, it has been proposed that granular tau oligomer is an intermediate form of tau fibrils [14]. Thus, monomeric tau molecules first bind to each other through disulfide bonds and SDS-resistant interactions to form tau oligomers that are not visible under AFM. Forty tau molecules bind together forming a β-sheet structure; these 40-tau aggregates, which appear granular in shape under AFM, accumulate to form tau fibrils.

2.6 Role of Granular Tau Oligomers in Neurodegeneration

Before forming fibrillar aggregates, tau forms two different types of tau aggregate: oligomeric tau (sarkosyl-soluble, not detectable by AFM) and granular oligomeric tau (sarkosyl-insoluble, detectable under AFM). If NFTs represent tombstones of neurodegeneration, these two kinds of intermediate tau oligomer may play a role in synapse loss, neuron loss, and ultimately, the neurodegeneration typically seen in tauopathies.

Kimura and colleagues [16] generated a transgenic (Tg) mouse line expressing P301L human tau. These mice display sarkosyl-insoluble tau aggregates and neuronal loss, but not typical NFTs. Because the pool of insoluble tau includes both fibrillar and granular forms of tau aggregates, and because NFTs are not a toxic species of tau [12–14], granular tau oligomers may be the form of tau involved in neuronal loss. This premise is in accord with the finding that the frontal cortex of

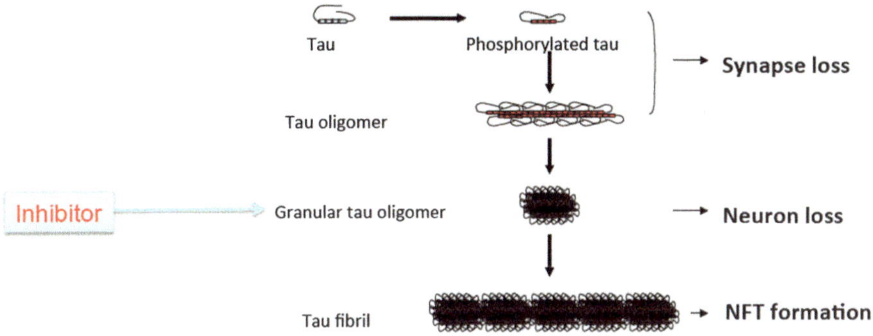

Fig. 2.1 Tau-targeting therapy. Before NFT formation, tau aggregates, and form tau oligomer, and granular tau oligomer. Hyperphosphorylated tau and tau oligomers are involved in synapse loss, and Granular tau oligomer causes neuronal loss. Blocking granular tau oligomer formation results in inhibition of neuronal loss accompanying ameliorate neuronal dysfunction in mouse model

Braak stage I brains contains elevated levels of granular tau, because the frontal lobe is particularly susceptible to volume decreases with aging [17].

Figure 2.1 summarizes the processes involved in tauopathy-associated neurodegeneration. As shown, hyperphosphorylated tau dislodges from microtubules, and since it has a higher affinity for participating in tau–tau interactions, individual hyperphosphorylated tau molecules bind to each other to form oligomeric tau; the latter is detergent-soluble and cannot be detected under AFM. Hyperphosphorylated tau or oligomeric tau is involved in synaptic loss, as demonstrated in studies of wild-type human tau Tg mice [16]. When the tau oligomer grows to about 40 molecules, it takes on a β-sheet structure and becomes a detergent-insoluble aggregate, visualized as a granular structure under AFM. Unlike hyperphosphorylated tau or oligomeric tau, granular tau oligomers may be involved in neuronal loss. As granular tau oligomers fuse together, they form tau fibrils, which ultimately form NFTs. Thus, different forms of tau aggregates are involved in the different pathological changes that occur in tauopathies. The mechanism underlying the loss of synapses and neurons seen in tauopathies may be the same mechanism underlying the neuropathology seen in other neurodegenerative diseases characterized by intracellular protein inclusions.

2.7 Tau-Targeting Therapy

As P301L-Tau mice do not form tau fibrils but nevertheless exhibit neuronal loss, we suggest that granular Tau is responsible for the toxicity of Tau aggregates. To test this notion further, we attempted to reduce formation of granular tau oligomers with the compound X1 which has the ability to inhibit the formation of granular tau. Interestingly, oral administration of X1 to P301L-Tau mice reduces neuronal loss, accompanied by a reduction of sarkosyl-insoluble tau levels, as compared to vehicle treatment. Taken together, our studies offer novel insights into Tau aggregation pathology; they strongly suggest that granular tau oligomers represent a toxic form of aggregated tau, and that X1 may be a promising compound for blocking the progression of AD.

References

1. Hardy J, Selkoe DJ (2002) The amyloid hypothesis of Alzheimer's disease: progress and problems on the road to therapeutics. Science 297(5580):353–356
2. Rosenblum WI (2014) Why Alzheimer trials fail: removing soluble oligomeric beta amyloid is essential, inconsistent, and difficult. Neurobiol Aging 35(5):969–974. doi:10.1016/j.neurobiolaging.2013.10.085
3. Jonsson T, Atwal JK, Steinberg S, Snaedal J, Jonsson PV, Bjornsson S, Stefansson H, Sulem P, Gudbjartsson D, Maloney J, Hoyte K, Gustafson A, Liu Y, Lu Y, Bhangale T, Graham RR, Huttenlocher J, Bjornsdottir G, Andreassen OA, Jonsson EG, Palotie A, Behrens TW, Magnusson OT, Kong A, Thorsteinsdottir U, Watts RJ, Stefansson K (2012) A mutation in APP protects against Alzheimer's disease and age-related cognitive decline. Nature 488 (7409):96–99. doi:10.1038/nature11283
4. Ingelsson M, Fukumoto H, Newell KL, Growdon JH, Hedley-Whyte ET, Frosch MP, Albert MS, Hyman BT, Irizarry MC (2004) Early Abeta accumulation and progressive synaptic loss, gliosis, and tangle formation in AD brain. Neurology 62(6):925–931
5. Gomez-Isla T, Hollister R, West H, Mui S, Growdon JH, Petersen RC, Parisi JE, Hyman BT (1997) Neuronal loss correlates with but exceeds neurofibrillary tangles in Alzheimer's disease. Ann Neurol 41(1):17–24
6. Goedert M, Spillantini MG (2000) Tau mutations in frontotemporal dementia FTDP-17 and their relevance for Alzheimer's disease. Biochim Biophys Acta 1502(1):110–121
7. Hutton M (2000) Molecular genetics of chromosome 17 tauopathies. Ann N Y Acad Sci 920:63–73
8. Spillantini MG, Van Swieten JC, Goedert M (2000) Tau gene mutations in frontotemporal dementia and parkinsonism linked to chromosome 17 (FTDP-17). Neurogenetics 2 (4):193–205
9. Braak H, Braak E (1996) Evolution of the neuropathology of Alzheimer's disease. Acta Neurol Scand Suppl 165:3–12
10. Samuel W, Masliah E, Hill LR, Butters N, Terry R (1994) Hippocampal connectivity and Alzheimer's dementia: effects of synapse loss and tangle frequency in a two-component model. Neurology 44(11):2081–2088
11. Ramsden M, Kotilinek L, Forster C, Paulson J, McGowan E, SantaCruz K, Guimaraes A, Yue M, Lewis J, Carlson G, Hutton M, Ashe KH (2005) Age-dependent neurofibrillary tangle formation, neuron loss, and memory impairment in a mouse model of human tauopathy (P301L). J Neurosci 25(46):10637–10647
12. Santacruz K, Lewis J, Spires T, Paulson J, Kotilinek L, Ingelsson M, Guimaraes A, DeTure M, Ramsden M, McGowan E, Forster C, Yue M, Orne J, Janus C, Mariash A, Kuskowski M, Hyman B, Hutton M, Ashe KH (2005) Tau suppression in a neurodegenerative mouse model improves memory function. Science 309(5733):476–481
13. Spires TL, Orne JD, SantaCruz K, Pitstick R, Carlson GA, Ashe KH, Hyman BT (2006) Region-specific dissociation of neuronal loss and neurofibrillary pathology in a mouse model of tauopathy. Am J Pathol 168(5):1598–1607

14. Maeda S, Sahara N, Saito Y, Murayama M, Yoshiike Y, Kim H, Miyasaka T, Murayama S, Ikai A, Takashima A (2007) Granular tau oligomers as intermediates of tau filaments. Biochemistry 46(12):3856–3861. doi:10.1021/bi061359o
15. Sahara N, Maeda S, Murayama M, Suzuki T, Dohmae N, Yen SH, Takashima A (2007) Assembly of two distinct dimers and higher-order oligomers from full-length tau. Eur J Neurosci 25(10):3020–3029
16. Kimura T, Yamashita S, Fukuda T, Park JM, Murayama M, Mizoroki T, Yoshiike Y, Sahara N, Takashima A (2007) Hyperphosphorylated tau in parahippocampal cortex impairs place learning in aged mice expressing wild-type human tau. EMBO J 26(24):5143–5152
17. Buckner RL (2004) Memory and executive function in aging and AD: multiple factors that cause decline and reserve factors that compensate. Neuron 44(1):195–208

Chapter 3
RNA Binding Proteins and the Genesis of Neurodegenerative Diseases

Benjamin Wolozin and Daniel Apicco

Recent advances in neurodegenerative diseases point to novel mechanisms of protein aggregation that revolve around the unique biology of RNA binding proteins. RNA binding proteins normally are present in the nucleus. Under conditions of cell stress these RNA binding proteins translocate to the cytoplasm where they form stress granules, which function in part to sequester specialized transcript and promote translation of protective proteins. Studies in humans show that pathological aggregates occurring in ALS, Alzheimer's disease, and other dementias co-localize with stress granules. One increasingly appealing hypothesis is that mutations in RNA binding proteins or prolonged periods of stress cause formation of very stable, pathological stress granules. The consolidation of RNA binding proteins away from the nucleus and neuronal arbors into pathological stress granules might impair the normal physiological activities of these RNA binding proteins causing the neurodegeneration associated with these diseases. Conversely, therapeutic strategies focusing on reducing formation of pathological stress granules might be neuroprotective.

We have recently identified a new type of molecular pathology in Alzheimer's disease (AD) that derives from the aggregation of RNA binding proteins (RBPs), forming RNA–protein complexes that include stress granules (SGs) [1]. These SGs progressively accumulate in the brains of transgenic models of tauopathy, as well as massively accumulate in subjects with AD and other neurodegenerative diseases [1, 2]. Some SGs (e.g., those positive for RNA binding protein TIA-1) co-localize with tau pathology, while other SGs (e.g., those positive for G3BP) often identify neurons that lack tau pathology. In order to understand how these RBPs might link to disease, it is useful to examine the unique biology of RBPs.

B. Wolozin, M.D., Ph.D. (✉) • D. Apicco
Departments of Pharmacology and Neurology, Boston University School of Medicine, Boston, MA, USA
e-mail: bwolozin@bu.edu

P. Vlamos, A. Alexiou (eds.), *GeNeDis 2014*, Advances in Experimental Medicine and Biology 822, DOI 10.1007/978-3-319-08927-0_3

mRNA binding proteins (RBPs) facilitate mRNA trafficking from the nucleus to the cytoplasm as part of the biological machinery that regulates mRNA metabolism, such as RNA translation, transport, and decay [3]. The functions of RBPs can generally be divided into nuclear and cytoplasmic activities, each of which constitutes two very large fields of literature. In the nucleus RBPs regulate mRNA maturation, including splicing, RNA helicase activity, RNA polymerase elongation, and nuclear export [4]. In the cytoplasm RBPs regulate RNA transport, silencing, translation, and degradation [3]. One of the most striking aspects of RBP function is their ability to reversibly aggregate to form RNA granules, which coalesce through binding of the low complexity, glycine-rich domains [5, 6]. Multiple different types of cytoplasmic granules exist, including transport granules, stress granules, and P-bodies. RNA degradation is mediated by a type of RNA granule, termed the P-body [7]. Transport granules move transcripts from the soma into the dendritic, and axonal arbors; axonal RBPs are more apparent during development and regeneration [8]. The presence of large numbers of RNA transport granules in neurons might explain why RBP dysfunction causes neuronal diseases. RBP complexes also mediate the process of activity-dependent protein synthesis, which is a critical element contributing to synaptic plasticity, habituation, and memory [9]. These granules interface with the microRNA system, since both microRNA and RBP regulate protein synthesis [10, 11]. The interaction of microRNA with RBP adds an additional layer of regulatory control.

Under stressful conditions mRNA binding proteins consolidate mRNA in cytoplasmic compartments, termed the stress granules (SGs) (Fig. 3.1); this recruitment is mediated by multiple proteins, including T-cell intracellular antigen 1 (TIA-1), RasGAP-associated endoribonuclease (G3BP), eukaryotic initiation factor 3 (eIF3), and poly-A binding protein (PABP) [12]. The process of SG formation is best understood for pathways mediated by phosphorylation of eIF2α. Stressful conditions prompt phosphorylation of eIF2α at serine 51, which inhibits formation of a complex containing eIF2, GTP, and tRNA [13]. Four different pathways lead to eIF2α phosphorylation: Oxidative stress (mediated by the kinase HRI), nutrient limitation or proteasomal dysfunction (mediated by the kinase GCN2), viral infection (mediated by the kinase PKR), and endoplasmic reticular stress (mediated by the kinase PERK) [14, 15]. SG formation is also regulated through independent pathways such as cleavage of tRNA by angiogenin or inhibition of ribosomal scanning [16, 17]. Stress also induces many RBPs to translocate from the nucleus to the cytoplasm. So, during stressful conditions, capped mRNA remains bound to the pre-initiation complex, which contains the other elongation factor binding proteins EF-4A, E, and G. This mRNA–protein complex is bound by SG-associated proteins, including eIF3, poly-A, and nucleating RBPs, such as TIA-1, TIAR, TTP, or G3BP [18, 19]. The SGs are initially small, but increase in size as the RBPs form cross-links by binding to each other through the glycine-rich protein aggregation domains.

An increasing number of mutations associated with neurodegenerative diseases are associated with RBPs. For instance, mutations in TDP-43, FUS, ataxin-2, matrin-2, hnRNPA1/B2, optineurin, and angiogenin are all linked to amyotrophic

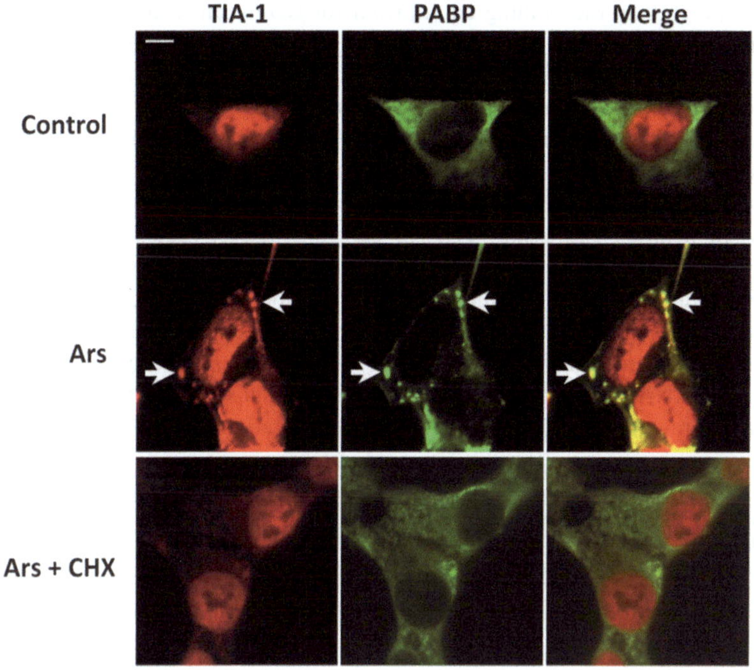

Fig. 3.1 Sodium arsenite induces stress granules co-positive for TIA-1 and PABP. Human SH-SY5Y cells (*top panels*, basal conditions) were treated with 0.5 mM sodium arsenite (Ars) for 30 min (*middle panels*) or arsenite plus 10 μg/ml cycloheximide (CHX, *lower panels*). *Arrowheads* denote stress granules co-positive for TIA-1 and PABP. Scale bar = 2 μm

lateral sclerosis [2, 20]. The interaction between pathology of SGs is easy to envision for RBPs that exhibit disease-linked mutations. Disease-linked mutations in these proteins generally increase the tendency of the protein to aggregate. Since these proteins normally exist in equilibrium between dispersed, soluble proteins and aggregated, insoluble complexes, aggregation-accelerating mutations would shift the equilibrium, leading to increased SG formation and formation of the stable, long-lived protein aggregates that we associate with disease pathology. The presence of more aggregates and/or aggregates with enhanced stability might also increase formation of secondary mature SGs around the aggregates, much like the normal evolution of SGs. This process of SG growth is analogous to cross-seeding, and normally forms the basis of SG maturation. Pathological cross-seeding has been shown in cell culture for polyglutamine-rich proteins, such as huntingtin [21]. This same process of accelerated cross-seeding might occur in the brain for other pathological proteins, more generally. The formation of persistent, insoluble protein aggregates might shift the equilibrium of RBPs towards aggregation, leading to accelerated, long-lived SG formation. These highly insoluble aggregates could also serve as a nidus for further aggregation of SGs, by cross-linking with other RBPs and also binding RNA as part of the process of SG maturation.

Thus, a shift in the binding equilibrium for RBPs can lead to excessive RNA granule formation. The shift leads to enhanced SG formation, with resulting formation of persistent SGs, which we hypothesize to be the substrate for much of the intracellular pathology that accumulates in neurodegenerative diseases. Recent studies by our group and that of the Taylor group reveal that mutations in RBPs, such as TDP-43, cause readily apparent changes even before pathology forms. We recently examined RNA transport granules, which travel along dendrites. In neurons expressing TDP-43 carrying disease-linked mutations, RNA transport granules are larger and move slower [22, 23]. This highlights mutation-related changes in RNA granules that could set the stage for subsequent disease. The innate biology build from reversible aggregation combined with the enhanced tendency of neurons to form cytoplasmic RNA granules (because of their strong need for mRNA transport) appears to provide the substrate for subsequent pathological aggregation and the resulting neurodegenerative disease.

Acknowledgments The research described above was supported by grants from NIEHS, Brightfocus foundation, Alzheimer Association, CurePSP foundation, and the Cure Alzheimer Foundation.

References

1. Vanderweyde T, Yu H, Varnum M, Liu-Yesucevitz L, Citro A, Ikezu T, Duff K, Wolozin B (2012) Contrasting pathology of stress granule proteins TIA-1 and G3BP in tauopathies. J Neurosci 32(24):8270–8283
2. Vanderweyde T, Youmans K, Liu-Yesucevitz L, Wolozin B (2013) Role of stress granules and RNA-binding proteins in neurodegeneration: a mini-review. Gerontology 59(6):524–533
3. Liu-Yesucevitz L, Bassell GJ, Gitler AD, Hart AC, Klann E, Richter JD, Warren ST, Wolozin B (2011) Local RNA translation at the synapse and in disease. J Neurosci 31(45):16086–16093
4. Heyd F, Lynch KW (2011) Degrade, move, regroup: signaling control of splicing proteins. Trends Biochem Sci 36(8):397–404
5. Han TW, Kato M, Xie S, Wu LC, Mirzaei H, Pei J, Chen M, Xie Y, Allen J, Xiao G, McKnight SL (2012) Cell-free formation of RNA granules: bound RNAs identify features and components of cellular assemblies. Cell 149(4):768–779
6. Kato M, Han TW, Xie S, Shi K, Du X, Wu LC, Mirzaei H, Goldsmith EJ, Longgood J, Pei J, Grishin NV, Frantz DE, Schneider JW, Chen S, Li L, Sawaya MR, Eisenberg D, Tycko R, McKnight SL (2012) Cell-free formation of RNA granules: low complexity sequence domains form dynamic fibers within hydrogels. Cell 149(4):753–767
7. Thomas MG, Loschi M, Desbats MA, Boccaccio GL (2011) RNA granules: the good, the bad and the ugly. Cell Signal 23(2):324–334
8. Krichevsky AM, Kosik KS (2001) Neuronal RNA granules: a link between RNA localization and stimulation-dependent translation. Neuron 32(4):683–696
9. Hoeffer CA, Klann E (2010) mTOR signaling: at the crossroads of plasticity, memory and disease. Trends Neurosci 33(2):67–75
10. Gibbings DJ, Ciaudo C, Erhardt M, Voinnet O (2009) Multivesicular bodies associate with components of miRNA effector complexes and modulate miRNA activity. Nat Cell Biol 11 (9):1143–1149
11. Schratt G (2009) microRNAs at the synapse. Nat Rev Neurosci 10(12):842–849

12. Anderson P, Kedersha N (2008) Stress granules: the Tao of RNA triage. Trends Biochem Sci 33(3):141–150
13. Kedersha NL, Gupta M, Li W, Miller I, Anderson P (1999) RNA-binding proteins TIA-1 and TIAR link the phosphorylation of eIF-2 alpha to the assembly of mammalian stress granules. J Cell Biol 147(7):1431–1442
14. Boyce M, Bryant KF, Jousse C, Long K, Harding HP, Scheuner D, Kaufman RJ, Ma D, Coen DM, Ron D, Yuan J (2005) A selective inhibitor of eIF2alpha dephosphorylation protects cells from ER stress. Science 307(5711):935–939
15. Mazroui R, Sukarieh R, Bordeleau ME, Kaufman RJ, Northcote P, Tanaka J, Gallouzi I, Pelletier J (2006) Inhibition of ribosome recruitment induces stress granule formation independently of eukaryotic initiation factor 2alpha phosphorylation. Mol Biol Cell 17 (10):4212–4219
16. Emara MM, Ivanov P, Hickman T, Dawra N, Tisdale S, Kedersha N, Hu GF, Anderson P (2010) Angiogenin-induced tRNA-derived stress-induced RNAs promote stress-induced stress granule assembly. J Biol Chem 285(14):10959–10968
17. Dang Y, Kedersha N, Low WK, Romo D, Gorospe M, Kaufman R, Anderson P, Liu JO (2006) Eukaryotic initiation factor 2alpha-independent pathway of stress granule induction by the natural product pateamine A. J Biol Chem 281(43):32870–32878
18. Wolozin B (2012) Regulated protein aggregation: stress granules and neurodegeneration. Mol Neurodegener 7:56
19. Kedersha N, Cho MR, Li W, Yacono PW, Chen S, Gilks N, Golan DE, Anderson P (2000) Dynamic shuttling of TIA-1 accompanies the recruitment of mRNA to mammalian stress granules. J Cell Biol 151(6):1257–1268
20. Johnson JO, Pioro EP, Boehringer A, Chia R, Feit H, Renton AE, Pliner HA, Abramzon Y, Marangi G, Winborn BJ, Gibbs JR, Nalls MA, Morgan S, Shoai M, Hardy J, Pittman A, Orrell RW, Malaspina A, Sidle KC, Fratta P, Harms MB, Baloh RH, Pestronk A, Weihl CC, Rogaeva E, Zinman L, Drory VE, Borghero G, Mora G, Calvo A, Rothstein JD, ITALSGEN Consortium, Drepper C, Sendtner M, Singleton AB, Taylor JP, Cookson MR, Restagno G, Sabatelli M, Bowser R, Chio A, Traynor BJ (2014) Mutations in the Matrin 3 gene cause familial amyotrophic lateral sclerosis. Nat Neurosci 17(5):664–666
21. Furukawa Y, Kaneko K, Matsumoto G, Kurosawa M, Nukina N (2009) Cross-seeding fibrillation of Q/N-rich proteins offers new pathomechanism of polyglutamine diseases. J Neurosci 29(16):5153–5162
22. Liu-Yesucevitz L, Lin AY, Ebata A, Boon JY, Reid W, Xu YF, Kobrin K, Murphy GJ, Petrucelli L, Wolozin B (2014) ALS-linked mutations enlarge TDP-43-enriched neuronal RNA granules in the dendritic arbor. J Neurosci 34(12):4167–4174
23. Alami NH, Smith RB, Carrasco MA, Williams LA, Winborn CS, Han SS, Kiskinis E, Winborn B, Freibaum BD, Kanagaraj A, Clare AJ, Badders NM, Bilican B, Chaum E, Chandran S, Shaw CE, Eggan KC, Maniatis T, Taylor JP (2014) Axonal transport of TDP-43 mRNA granules is impaired by ALS-causing mutations. Neuron 81(3):536–543

Chapter 4
Individual Severity of AD-Type Lesions, Aß Oligomers, and Comorbidity in the Brain Aging

Constantin Bouras, Eniko Kovari, Armand Savioz, Ezio Giacobini, and Aikaterini Xekardaki

In cognitively intact individuals and patients with Alzheimer's disease (AD) the formation of neurofibrillary tangles (NFTs), senile plaques (SPs), and the synaptic loss characterizes the neuropathology of brain aging. There is a differential cortical vulnerability to the degenerative process in extreme brain aging. In the oldest-old population the distribution and the severity of NFTs and SPs could be different compared to younger persons.

It is well known, actually, that the proliferation of NFTs in the temporal neocortex is correlated with clinical signs of dementia. On the other hand, in normal aging NFTs are usually restricted to the hippocampus.

The toxicity of Aβ oligomers has been postulated after the absence of correlation between SPs and the clinical severity. The distribution of Aß oligomers, amyloid plaques, and NFTs has been studied, with immunohistochemical methods, in the hippocampal formation and the temporal neocortex as well as in the frontal and occipital cortical areas, in 74 cases (43–104 years old). In all these anatomical areas a parallel evolution in the presence and the severity of Aß oligomers and SPs has been observed. Moreover, the study by western blots for the small oligomers, comparing normal aging and AD, will be reported.

Moreover, in aging brain the important comorbidity mainly concerning the degenerative pathology and the vascular pathology is also reported (neuropathological material of 10,000 cases) as well as the progressive accumulation of vascular lesions, including the amyloid angiopathy, throughout the human life span.

C. Bouras (✉) • E. Kovari • A. Savioz • A. Xekardaki
Department of Psychiatry, University Hospitals, Geneva, Switzerland
e-mail: Constantin.Bouras@unige.ch; eniko.kovari@hcuge.ch; Armand.Savioz@hcuge.ch; Aikaterini.Xekardaki@hcuge.ch

E. Giacobini
Department of Internal Medicine, Rehabilitation and Geriatrics, University Hospitals, Geneva, Switzerland
e-mail: Ezio.Giacobini@hcuge.ch

P. Vlamos, A. Alexiou (eds.), *GeNeDis 2014*, Advances in Experimental Medicine and Biology 822, DOI 10.1007/978-3-319-08927-0_4

Chapter 5
The Influence of Immunomodulatory Treatment on the Clinical Course of Multiple Sclerosis

Andrius Kavaliunas, Leszek Stawiarz, Jonas Hedbom, Anna Glaser, and Jan Hillert

Abstract *Background*. Multiple sclerosis (MS) is a chronic disease of the central nervous system. One of the major questions concerning the clinical progression of MS, still insufficiently elaborated or confirmed, is if it can be slowed down or augmented by external factors. Immunomodulatory treatment is a disease modifiable factor shown to influence disease progression of various medical conditions.

Objective. To investigate if treatment affects the long-term clinical progression of MS, measured as time from diagnosis to score of 4 or higher of Expanded Disability Status Scale (EDSS).

Methods. Longitudinal, prospective data concerning treatment status and EDSS were collected by health professionals in the Swedish MS Registry. Study cohort comprised new diagnosed MS patients at Karolinska Hospital between 2001 and 2005. Survival analysis adjusted for suspected confounders was used with the outcome variable time from diagnosis to EDSS $\geq$4.

Results. Early treatment was correlated with longer time from diagnosis to EDSS $\geq$4 (HR: 1.77; 95 % CI: 1.15–2.73; $p = 0.01$). Additionally, the influence of the covariates—age at onset and the baseline EDSS, which were statistically significant with hazard ratios of 1.03 and 2.1, respectively, was found.

Conclusion. Early treatment was associated with a better clinical outcome.

Keywords Multiple sclerosis • Disease modifying drugs • Disease progression • Survival analysis

5.1 Introduction

Multiple sclerosis (MS) is a chronic inflammatory disease of the central nervous system that implies impaired motor and cognitive functions. It is characterized by fluctuations in both disease severity and occurrence of symptoms at different

A. Kavaliunas (✉) • L. Stawiarz • J. Hedbom • A. Glaser • J. Hillert
Department of Clinical Neuroscience, Karolinska Institute, Stockholm, Sweden
e-mail: andrius.kavaliunas@ki.se

P. Vlamos, A. Alexiou (eds.), *GeNeDis 2014*, Advances in Experimental Medicine and Biology 822, DOI 10.1007/978-3-319-08927-0_5

periods during MS progression of each individual patient. The early clinical courses of MS are relapsing-remitting MS (RRMS) and primary progressive MS (PPMS). A relapse is an acute inflammatory event where the patient suddenly experiences a period of significantly worse neurological impairment that typically completely resolves after some time. Patients with RRMS usually convert into a progressive course with or without superimposed exacerbations called secondary progressive MS (SPMS). The annual conversion rate into SPMS is approximately 2.5 % [1]. Within 20–30 years the majority of RRMS patients converts to SPMS [2].

Within MS population the spectrum of patients ranges from essentially unaffected to completely disabled. The grade of disability is routinely quantified by neurologists according to the Expanded Disability Status Scale (EDSS) [3]. The EDSS spans between 0 and 10 with increments of 0.5 where 0 is a neurologically unaffected patients and 10 is death by MS. It is agreed among physicians and researchers that EDSS scores of 4 and 6 are used as main disease milestones. The time it takes to reach such a milestone can be viewed as a proxy parameter of progression.

Disease-modifying drugs (DMDs) for MS have been available since the mid-1990s. Most randomized control trials, which showed beneficial responses to DMDs, have a follow-up time of 2 years, which is too short to answer questions concerning long-term progression in MS. It has therefore been of interest to evaluate if therapy had effect on the clinical outcome with longer follow-ups. A general idea has been that patients benefit the most from the early treatment and this could be beneficial to patient's long-time prognosis [4]. The theory behind this idea seems to be more of an intuitive nature than purely mechanisms oriented, and some researchers strongly oppose it [5]. The confusion surrounding best practice put physicians in a difficult position when choosing: if and when to treat. Therapy with DMDs includes the risk of irreversible side effects; moreover, it is expensive.

The aim of the study was to investigate if treatment, measured as time from disease onset to score of 4 or higher of EDSS, affected the clinical disease progression of MS.

5.2 Methods

This primarily prospective study investigated MS patients from an inception cohort, StopMS, initiated 2001-01-01. The study had a 10 years follow-up. Within the study, information concerning treatment, disability, relapses, and progression was gathered and put into the Swedish Multiple Sclerosis Registry at every health care visit [6]. Inclusion criteria: MS diagnosis during 2001–2005 at Karolinska University Hospital (Solna and Huddinge). Exclusion criteria: not given consent to the study. Required ethical permission has been obtained.

Six hundred and twenty-four patients were included in a cohort. There was information concerning 4,240 EDSS evaluations on 593 patients in the dataset, all done by neurologists. If a patient was under relapse at EDSS scoring, this particular evaluation was excluded from the analysis (three patients were excluded due to this). EDSS variables were the baseline EDSS (the first that was not recorded during the relapse) and the time to EDSS 4 or higher (or censored case) in months. There were 203 events and 467 censored cases. Treatment information included 1,316 prescribed drugs on a total of 504 patients. Time to treatment (TtT)—the time from MS onset to the first active treatment with a DMD was calculated. The patients were divided into two groups according to TtT: early or delayed.

All DMD treatments were taken to increase the amount of patients in the analysis. The choice to use survival analysis (Kaplan-Meier and Cox proportional hazard) was motivated by possibility to increase the amount of information that could be analyzed by including censored cases (patients that never reached EDSS 4 during the follow-up). Kaplan-Meier comparison between the groups was performed with log-rank significance test. Baseline EDSS, age at onset, and gender were analyzed as confounders using Cox proportional hazard model. The statistical software SPSS (version 22) was used for the analysis. Differences were defined as statistically significant for p-values lower than 0.05.

5.3 Results

The baseline characteristics of the patients are presented in Table 5.1. The patients were divided into two groups according to TtT: early treatment (205 patients that received the first treatment in 24 months from MS onset) and delayed treatment (299 patients that received the first treatment after 24 months from MS onset). These two groups did not differ considerably in their baseline characteristics.

We performed univariate Kaplan-Meier analysis with time from diagnosis to EDSS ≥ 4 as the outcome measure. There were 161 events out of 490 cases. The initial analysis is presented in Fig. 5.1. The results show statistically significant difference for the time from onset to reach EDSS 4 between early and delayed treatment groups ($p > 0.001$).

However a distinct drop in the very first months can be observed in the figure. Therefore we excluded 58 cases which had EDSS 4 or higher at baseline in further analysis.

Difference in time from onset to reach EDSS 4 remained statistically significant after adjusting for covariates as seen in Table 5.2. Patients who received delayed treatment reached EDSS 4 sooner with hazard ratio of 1.77 (95 % CI: 1.15–2.73). Additionally, age at onset and the baseline EDSS were statistically significant with hazard ratios of 1.03 and 2.1, respectively. The Kaplan-Meier survival curves for the early versus delayed treatment groups at mean of covariates are shown in Fig. 5.2.

Table 5.1 Baseline characteristics for early and delayed treatment groups

Variable	TtT ($n = 504$) Early ($n = 205$) Count (%) Mean (95 % CI)	Delayed ($n = 299$) Count (%) Mean (95 % CI)
Gender:		
Female	142 (69 %)	227 (76 %)
Male	63 (31 %)	72 (24 %)
Baseline EDSS	1.6 (1.4–1.7)	2.2 (2.0–2.4)
Age at onset (years)	32.0 (30.7–33.4)	31.4 (30.2–32.5)
TtT (months)	9.7 (8.8–10.5)	104.4 (95.2–113.7)

TtT Time from MS onset to the first active treatment with a DMD

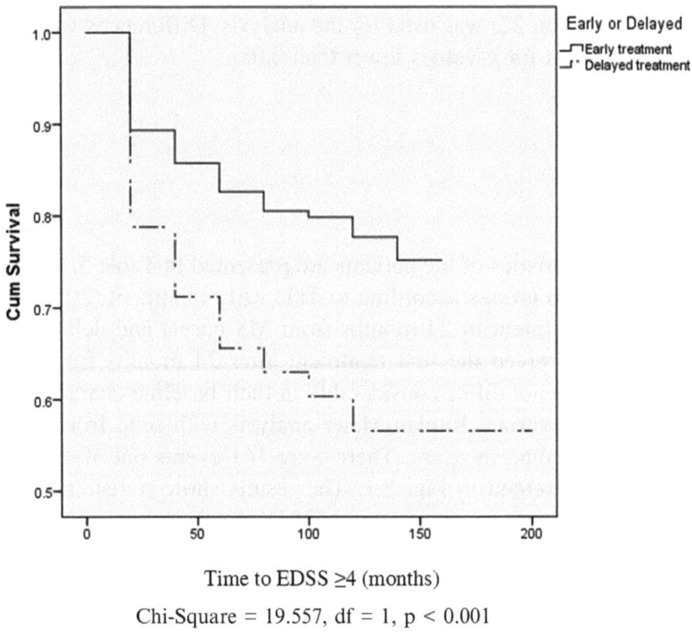

Fig. 5.1 Kaplan-Meier survival curves for early and delayed treatment groups

Table 5.2 Hazard ratios for time from MS onset to EDSS 4 of early and delayed treatment groups

Covariate	B	SE	Wald	p-value	Hazard ratio	95 % CI for HR Lower	95 % CI for HR Upper
TtT	0.570	0.222	6.596	0.01	1.769	1.145	2.734
Baseline EDSS	0.742	0.111	45.000	<0.001	2.101	1.691	2.609
Age at onset	0.029	0.011	7.197	0.007	1.029	1.008	1.051
Gender	0.01	0.217	0.002	0.964	1.01	0.660	1.546

TtT Time from MS onset to the first active treatment with a DMD, *B* intercept, *SE* standard error, *Wald* Wald chi-square test

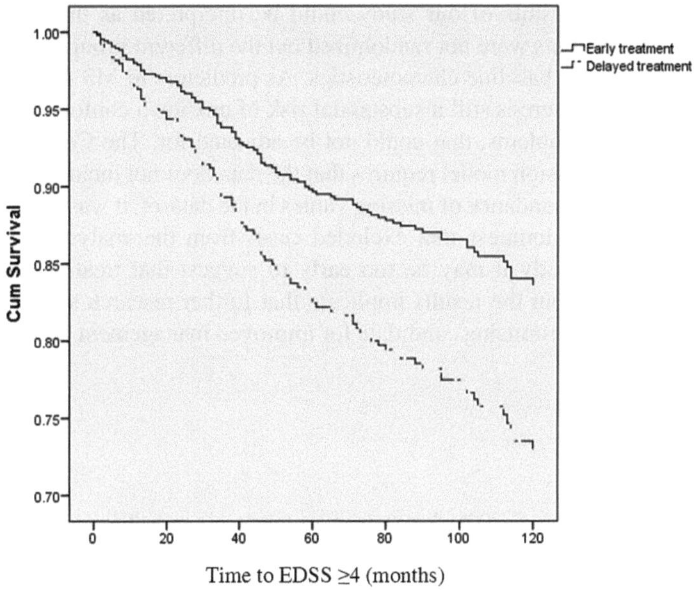

Early	184	176	167	159	146	94	51
Delayed	248	230	205	184	162	118	69

Fig. 5.2 Kaplan-Meier survival curves at mean of covariates for early and delayed treatment groups after adjusting for covariates

5.4 Discussion

We investigated treatment effect on MS prognosis, using a unique material of prospectively included patients in a real-world setting of an MS clinic. Our data indicate that early treatment in contrast to delayed treatment may be beneficial in slowing MS progression within 10 years. As it appeared in the results, the delayed treatment group had an almost 80 % increased risk of reaching EDSS 4. This study supports for treatment, and especially early treatment. We have found that baseline EDSS and age at onset were meaningful confounders. The fact that age at onset was significant was well in line with previous research where age at onset almost always was the best predictor of the rate of disease progression [1, 7] and higher baseline EDSS score's association with a higher hazard of reaching and EDSS score of 6 [8]. The age at onset was highly significant covariate in time to SP for patients with RRMS [9].

There are few studies that have evaluated the effects of early and delayed treatment on progression in MS. It was found that the early treatment reduced the risk of transformation of clinically isolated syndrome (CIS) into RRMS in the 3-year follow-up [10]. After 5 years follow-up the outcome was EDSS progression and early treatment did have better results but these were not significant [11].

Essentially, the results of our study could be interpreted as that treatment is beneficial. The patients were not randomized but the different groups did not differ considerably in their baseline characteristics. As predictors in MS progression are poorly understood, there is still a substantial risk of unknown confounding factors, such as cognition problems, that could not be adjusted for. The Cox proportional hazard as each regression model requires that the data does not include any missing values. Because of abundance of missing values in the dataset, it was not possible to control apparent randomness that excluded cases from the analysis. Because of limitations in this study it may be too early to suggest that treatment strategies should be changed, but the results implicate that further research should evaluate early treatment as a promising candidate for improved management of MS patients.

References

1. Confavreux C, Vukusic S (2006) Age at disability milestones in multiple sclerosis. Brain 129 (Pt 3):595–605
2. Compston A, Coles A (2008) Multiple sclerosis. Lancet 372(9648):1502–1517
3. Kurtzke JF (1983) Rating neurologic impairment in multiple sclerosis: an expanded disability status scale (EDSS). Neurology 33(11):1444–1452
4. Elovaara I (2011) Early treatment in multiple sclerosis. J Neurol Sci 311:S24–S28
5. Gilmore CP, Cottrell DA, Scolding NJ, Wingerchuk DM, Weinshenker BG, Boggild M (2010) A window of opportunity for no treatment in early multiple sclerosis? Mult Scler 16 (6):756–759
6. Andersen O (2012) From the Gothenburg cohort to the Swedish multiple sclerosis registry. Acta Neurol Scand Suppl 195:13–19
7. Langer-Gould A, Popat RA, Huang SM, Cobb K, Fontoura P, Gould MK, Nelson LM (2006) Clinical and demographic predictors of long-term disability in patients with relapsing-remitting multiple sclerosis: a systematic review. Arch Neurol 63(12):1686–1691
8. Shirani A, Zhao Y, Karim ME, Evans C, Kingwell E, van der Kop ML, Oger J, Gustafson P, Petkau J, Tremlett H (2012) Association between use of interferon beta and progression of disability in patients with relapsing-remitting multiple sclerosis. JAMA 308(3):247–256
9. Tedeholm H, Lycke J, Skoog B, Lisovskaja V, Hillert J, Dahle C, Fagius J, Fredrikson S, Landtblom AM, Malmeström C, Martin C, Piehl F, Runmarker B, Stawiarz L, Vrethem M, Nerman O, Andersen O (2013) Time to secondary progression in patients with multiple sclerosis who were treated with first generation immunomodulating drugs. Mult Scler 19 (6):765–774
10. Kappos L, Freedman MS, Polman CH, Edan G, Hartung HP, Miller DH, Montalbán X, Barkhof F, Radü EW, Bauer L, Dahms S, Lanius V, Pohl C, Sandbrink R, BENEFIT Study Group (2007) Effect of early versus delayed interferon beta-1b treatment on disability after a first clinical event suggestive of multiple sclerosis: a 3-year follow-up analysis of the BENEFIT study. Lancet 370(9585):389–397
11. Kappos L, Freedman MS, Polman CH, Edan G, Hartung HP, Miller DH, Montalbán X, Barkhof F, Radü EW, Metzig C, Bauer L, Lanius V, Sandbrink R, Pohl C, BENEFIT Study Group (2009) Long-term effect of early treatment with interferon beta-1b after a first clinical event suggestive of multiple sclerosis: 5-year active treatment extension of the phase 3 BENEFIT trial. Lancet Neurol 8(11):987–997

Chapter 6
Notch Signaling and Ageing

Eleftheria Polychronidou, Dimitrios Vlachakis, Panayiotis Vlamos, Marc Baumann, and Sophia Kossida

Abstract Notch signaling is a master controller of the neural stem cell and neural development maintaining a significant role in the normal brain function. Notch genes are involved in embryogenesis, nervous system, and cardiovascular and endocrine function. On the other side, there are studies representing the involvement of Notch mutations in sporadic Alzheimer disease, other neurodegenerative diseases such as Down syndrome, Pick's and Prion's disease, and CADASIL. This manuscript attempts to present a holistic view of the positive or negative contribution of Notch signaling in the adult brain, and at the same time to present and promote the promising research fields of study.

6.1 Introduction

The normal brain development is a complicated task and is combining the cooperation of vital functions. Notch signaling is controlling the neural stem cell and development [1] and maintains a significant role in the normal brain function. A protein that is related to brain function is the Notch protein and is connected with vertebrate and invertebrate species. Notch possesses four receptors that were duplicated several times leading to four genes in mammals NOTCH1, NOTCH2, NOTCH3, NOTCH4 [2] and many ligands, including jagged 1 (JAG1) and JAG2 (homologues of serrate), and delta-like proteins. Notch and its ligands are single-pass transmembrane heterodimers. They are characterized by the repeated extracellular EGF (epidermal growth factor) domains and the notch domains; they have

E. Polychronidou • D. Vlachakis • S. Kossida (✉)
Bioinformatics and Medical Informatics Team, Biomedical Research Foundation, Academy of Athens, Soranou Efessiou 4, Athens 11527, Greece
e-mail: skossida@bioacademy.gr

P. Vlamos
Department of Informatics, Ionian University, Corfu, Greece

M. Baumann
Protein Chemistry/Proteomics Unit, Biomedicum Helsinki, Institute of Helsinki, P.O. Box 36, Helsinki, Finland

P. Vlamos, A. Alexiou (eds.), *GeNeDis 2014*, Advances in Experimental Medicine and Biology 822, DOI 10.1007/978-3-319-08927-0_6

the capacity to increase the gene expression by binding to an activator. Notch can be characterized as a key player in several development processes, maintaining an instructive role in the cell–cell communication regulating important processes.

Particularly, Notch genes are involved in embryogenesis, as is required in embryo polarity and in somitogenesis [3, 4]. Notch signaling is central for the nervous system operation and specifically in the cell differentiation, neurite development, gliogenesis, and brain function of the elderly [5–8]. In cardiovascular development, Notch signal pathway focuses on the process of the atrioventricular canal, the myocardial, and the cardiac outflow track [9]. Notch signaling is also shown to be necessary in the endocrine system [10], and especially pancreatic function, intestinal development [11], and the bone development [12].

In 1914, the Notch gene was noticed in the wing blades of the fly *Drosophila melanogaster* [13]. A number of alternative forms of the same gene were identified by Thomas Hunt Morgan in 1917 [14]. In 1940, Poulson underlined the relationship between the genomic locus of Notch gene and the embryonic development [6]. The first attempt to analyze the molecule was undertaken by Spyros Artavanis-Tsakonas and Michael W. Young in the late 1980s. A later genomic analysis in Notch area introduced the involvement of Notch, Lin-12, Glp-1 receptor family in the development of *Drosophila melanogaster* and *Caenorhabditis elegans* [3, 15, 16].

According to literature, Notch homologues seem to originate from the gene duplication in vertebrate and invertebrate species. In particular, from the syntenic region of human Notch 1, 2, and 3 became clear that there was double gene duplication during metazoan evolution making sufficient their necessity [17]. On the other hand, in organisms less complex than Drosophila, the Notch genes are emerging from independent duplications. Example of this case represents *Caenorhabditis elegans* [18]. In 2001, Kortschak described that Notch1a and Notch1b are results from duplication near teleost/mammalian divergence and that Notch4 is a result of a Notch3 divergence [17]. However, there is a doubt about the origin of this duplication [17, 18], which is limiting the extensive investigation on Notch progression. A potential solution on filling the information gap of the Notch origin could constitute the analytical study of other genomes, like the amphioxus genome whose draft assembly genome was identified during the last years [19].

Lately, there studies placing Notch action as a master regulator of plasticity in the adult brain—from stem cells to mature neurons to degenerating neurons. The findings so far set the foundation for further structural, functional, and evolutionary investigations, although many queries still can be formulated on the validity of the information, focusing on the expansion and coevolution of the Notch genes [2].

6.2 Notch in Adulthood

Notch proteins held a key role in nervous system and brain functions especially when this is linked to elderly. Summarizing the contribution of the Notch signaling pathway in adulthood, it is important to underline that represents a key role in the nervous system development and brain functions [20]. Due to its activity, in the central nervous system, it can maintain functional plasticity throughout life and experimental pieces of evidence suggesting that Notch is operative in diverse brain pathologies including tumorigenesis, stroke, and neurological disorders such as Alzheimer's disease, Down syndrome, and multiple sclerosis [21].

More particularly, Notch1 is expressed in pyramid neurons of the cortex and hippocampus providing synaptic plasticity and memory processing and regulating spine morphology [5, 22]. Despite that fact, Notch1 is strongly induced following brain injury and may contribute to cell death [23]. The expression of Notch2 is very similar to that of Notch1 and is upregulated upon injury, contributing to brain damage [24]. Notch3 is mostly expressed in astroglial progenitors, choroid plexus, and vasculature, whereas Notch4 has not been detected so far [22].

The function of Notch is clarified through the causes of its mutations. More specifically, Notch signaling pathway element mutation is involved in Alagille [25], in CADASIL (cerebral autosomal dominant arteriopathy with subcortical infarcts and leukoencephalopathy) [26], and in Hajdu–Cheney syndromes [27] and abnormal Notch expression levels are related with Down syndrome [28] and Alzheimer disease [29]. For example, Alagille patients carrying genetic mutation for Jagged1, except the liver and the heart pathology, also display signs of mental retardation related to Notch [22].

The research into Notch functions could assert the aforementioned scenarios as in the study of the specific mutations in the human gene encoding lamin A or in the lamin A-processing enzyme, Zmpste24, that causes premature ageing. According to the authors, new data on mice and humans suggest that these mutations affect adult stem cells by interfering with the Notch signaling pathways [30]. Another study which emphasizes in the significant contribution of Notch signaling is their relation with the regeneration of muscles. Conboy et al. in 2003 analyzed the Notch signaling focusing on its important determinant of muscle regenerative potential that declines with age. Costa et al. proved experimentally in 2003 that adult mice heterozygous for mutations in either Notch1 or Cbf1 have deficits in spatial learning and memory [5]. Another case study proposes Notch signaling as the new potential target for microtubule stabilization [31]. As far is concerned neurodegenerative diseases are connected with cytoskeletal dysfunction. The key element of the normal cytoskeletal function is the microtubules and their dysregulation which can cause axonal transport impairment, synaptic contact degeneration, and impaired neuronal function leading finally to neuronal loss. Notch signaling pathway fills the requirements of being the most important regulator in order to increase cytoskeleton plasticity with intense neurite remodeling.

Clinical trials in Alzheimer present a statistically significant worsening of cognition relative to controls when Notch signaling is spotted [32]. The last study accomplished in 2011 focused on the investigation of the frequency of Notch3 variations in community-dwelling elderly and their effect on cerebral small vessel disease related magnetic resonance imaging phenotypes. Attempting to describe the function of the Notch receptors and ligands in mammalian brain, there are still many unclarified parts. In a study carried out in 2001, the expression patterns for three receptors of this system, Notch1, 2, and 3, in late embryonic and postnatal rat brain were examined by in situ hybridization. The results presented that both Notch1 and Notch3 mRNAs are expressed along the inner aspect of the dentate gyrus, a site of adult neurogenesis [33].

The contribution of Notch signaling in ageing as it was described in the paragraphs above is connected with the conservation of the adult nervous system functions. Recently, the effort of studying the relation between Notch proteins and degenerative disorders which are mainly related to ageing is becoming more and more widespread resulting in promising treatment possibilities. A field of study relative to this assumptions consists the study of relation between Notch signaling and Alzheimer disease.

6.3 Notch and Alzheimer

Alzheimer's disease (AD) is a widespread disease and the most common cause of dementia in the elderly. It is affecting over 35 million people worldwide (Alzheimer's Association 2009). Alzheimer's dementia is the result of the uncoupling of the neurotransmission machinery from the neurovascular blood supply chain [29]. The primary pathogenic mechanism of dementia is the deposits of amyloid-beta (Aβ) peptide in the center of the senile plaques. The following stage in the mechanism is characterized by the initial β-secretase cleavage of the amyloid precursor protein in the extracellular domain resulting in a transmembrane cleavage of a C99 intermediate by γ-secretase. Although the amyloid hypothesis constitutes the central scenario of Alzheimer's pathology, still the nonpathological function of amyloid precursor protein remains unknown, creating an area of research.

The Notch signaling was found to be critical mainly for the nervous system development and function. The connection of Alzheimer disease—especially Aβ levels—and Notch signaling was realized through the clinical trials of amyloid hypothesis. According to amyloid hypothesis, the blocking of secretases should lower the Aβ levels, benefiting AD patients. The results proved to be disappointing, as the Aβ suppression induced the same result in Notch signaling, which suppresses the EphB receptor expression by neurons, maintaining dense blood vessels branching [34]. Despite this case, the β-secretase inhibitors should decrease the Aβ levels, without inhibiting Notch. Eventually, the β-secretase inhibitors cause general weakness to the patients, results that prove to be the same in mice cases after excessive Notch signaling. Summarizing, if the amyloid-Notch hypothesis proved to be correct, the patients treated with β-secretase inhibitors may show improved memory but less synaptic plasticity [34].

By the moment, there are two relative studies representing the significant increase in Notch expression activity in sporadic Alzheimer disease [35, 36]. Additionally, other neurodegenerative diseases such as Down syndrome [28], Pick's [36] and Prion's disease [37] are depicted by the increase in Notch expression and amyloid plaques formation. The area that needs to be further investigated concerns the reason that progressive neurodegenerative diseases have different mechanisms in terms of Notch processing and signaling, and whether Notch effectively contributes to their pathobiology.

Another scenario that links the Alzheimer disease with the functions of Notch proteins is the damage they produce in small blood vessels [29]. Late studies of living patients proved that lesions in white matter regions of the brain lead to the appearance of amyloid deposits and are related with damaged small blood vessels. According to literature Notch3 mutations damage small blood vessels of white matter regions and cause dementia. Since both types of dementia are related to similar neuropathological findings there is a possibility of similar pathogenesis [38].

6.4 Focusing on Notch3

Notch3 is a protein and the third out of four homologues of the *Drosophila melanogaster* [13] type I membrane protein notch. The protein Notch3 is essential for the maintenance of normal and healthy muscle cell activity in the arteries of the brain. Following the function of the gene, after the production of Notch3 receptors, they send signals to the nucleus of the cell when the appropriate molecule binds to the receptor. The signals activate genes within vascular smooth muscle cells. The Notch3 gene is located on the short (p) arm of chromosome 19 between positions 13.2 and 13.1. It consists of 33 exons spanning 7 kilobases and encodes a transmembrane receptor of 2,321 amino acids involved in cellular signaling and fate during embryonic development [39].

Notch3 signaling pathway can be treated as an important variable in the identification of cancer cells like breast, lungs, ovarian, liver, and other types of tumor cells. A typical example of this function is that Notch3 has a crucial role in the proliferation of ErbB2-Negative human breast cancer cells [40]. Li et al. [41] described the importance of NOTCH3–HES-5 signaling pathway in the development of pulmonary arterial hypertension and at the same time they suggested it as a target pathway for therapeutic intervention [41]. According to Alunni et al. [42] Notch3 signaling gates cell cycle entry and limits neural stem cell amplification in the adult pallium. In this article, the authors implicate that Notch3 is at the top of this hierarchy to gate neural stem cell activation and amplification, protecting the homeostasis of adult neural stem cell reservoirs under physiological conditions.

There have been recorded more than 130 Notch3 gene mutations connected with cerebral autosomal dominant arteriopathy with subcortical infarcts and leukoencephalopathy, commonly known as CADASIL [43]. The mutations of Notch3 mainly focus on the addition or deletion of one protein building block (amino acid) in the Notch3 receptor, the cysteine, which affects the vascular smooth muscle cell causing CADASIL symptoms. A study accomplished in 2011 investigate the frequency of Notch3 variations in community-dwelling elderly and their effect on cerebral small vessel disease related magnetic resonance imaging phenotypes [44]. The results presented that Notch3 is highly variable with both common and rare single nucleotide polymorphisms spreading across the gene, and that common variants at the Notch3 gene increase the risk of age-related white matter lesions in hypertensives. In 2012, Guerreiro et al. detect through an exome sequence a connection between a Notch3 mutation and Alzheimer disease [32], focusing on the ability of exome sequencing in the efficient genetic mutations identification.

6.5 Notch3 and CADASIL

As it was described above, the Notch3 is a receptor protein that is strategically positioned on the surface of smooth muscle cell very close to the local blood vessels. Following Poulson's findings it was found that if Notch switches off and becomes inactive, epidermal precursors kick in that convert normal cells to neuroblasts [45]. Neuroblasts differentiate and produce embryos that have nervous system hypertrophy and epidermal structure deficiencies. Bill Welshon then studied the 3C7 region on the X chromosome, where Notch is located. It was then that a very detailed map of Notch mutations was first drawn. In flies Notch is affecting a series of biological characteristics, including the definition of boundaries between cells with developmental roles. Overall it is well established that Notch activity influences differentiation, proliferation, and apoptosis.

Notch has been linked mainly to three inherited diseases, one of which is CADASIL. An excess of 190 mutations of the Notch3 gene has been found to induce CADASIL. CADASIL is a hereditary disease affecting over middle-aged adults, leading them to disability and dementia. CADASIL stands for cerebral autosomal dominant arteriopathy with subcortical infarcts and leukoencephalopathy. It was first identified as a disease in 1993, even though there are records pointing back to 1955, when van Bogaert characterized CADASIL as the Binswanger disease [45]. After a series of other patient incidences, CADASIL disease was linked to the Notch homolog 3 protein. The actual prevalence of the disease is unknown. However, CADASIL has been reported in more than 500 families around the globe. The clinical manifestation of CADASIL can be described by five distinct symptoms. Those are migraine with aura, subcortical ischemic events, mood disturbance, apathy, and cognitive impairment. The symptoms can vary depending on patient age and progression of the disease [45].

Initiation of the Notch3 function and activation is triggered by small molecular ligands that bind to certain areas of the Notch3 receptor and activate it. The trans interactions between Notch and Delta are well documented [46]. However, it is the ratio of ligand and receptor adjacent cell expression that dictate the level of activation. Each one of those two adjacent cells will have to take active roles as either a signal receiving or a signal sending cell. The mechanism behind this is the critical ratio levels between functional ligand and receptor in each cell. The signal sending cell will be the one expressing more ligand whereas the signal receiving cell will be the one expressing more receptor [46]. This asymmetry is mediated by feedback loops, whose mechanism still remains to be elucidated.

As mentioned above, Notch is capable of influencing cell proliferation, differentiation, and apoptosis. Thereby, it is directly linked to major cell anomalies such as cancer. Notch activation will definitely affect cell proliferation in a repertoire of cellular contexts. It is noteworthy that activation of Notch can even influence the proliferation patterns of nonadjacent cells. Today, the exact molecular and structural mechanism behind this phenomenon remains elusive. For example, it has been demonstrated in a mouse tumor model that Notch activation in the crypts of the adult mouse led to dramatic proliferation increase in the area that is responsible for keeping the stem cells that are used for the programmed regeneration of the intestine.

Another example involves the vascular smooth muscle cells network. Notch3 is very important when it comes to the survival or death of the vascular smooth muscle cell. Not surprisingly the Notch3 receptors are key proteins that play a pivotal role in the optimal functionality, survival, and maintenance of muscle cells in the arterial network of the brain. The most involved amino acid is cysteine. It is either the loss or the addition of a cysteine residue in the EGF-like domain of Notch3 protein that alters the molecular properties of the protein in vascular smooth muscle cells. The vascular smooth muscle cell damage is inflicted via a series of recurring strokes. Overall, one can say that it is not Notch itself that is causing the tumor in the intestine, but it is surely somehow involved in the increased cell population that may lead to cancer under the right conditions [47].

At the moment there is no treatment available for CADASIL and thereof no drug that can act specifically on the Notch3 protein receptor. Medical practitioners prescribe aspirin, dipyridamole, or clopidogrel, which are found to limit the symptoms of the disease and to relatively slow it down.

6.6 3D Molecular Modeling Study on the Human NOTCH EGFS Protein

The homology modeling of the Notch3 protein was carried out using a restraint-based approach, as implemented in the MODELLER package [48]. The crystal structure of the HUMAN NOTCH-1 EGFS 11-13 (PDB entry: 2VJ3) [49] was used as template structure. The sequence alignment between the target sequence of TbTOP1B and the template sequence revealed almost 69 % identity, which allowed for a reliable homology modeling to be performed. The homology modeling process in Modeller starts with the building of a profile using the "build_profile.py" script. Then the alignment was performed using the "align2d" command. As soon as the alignment between the target and the template was constructed, the 3D model of the target was calculated using the automodel class of Modeller. Finally, the model was evaluated using the DOPE approach via the "evaluate_model.py" file. The overall homology modeling process was divided into the following steps: First, the initial spatial constraints for the target sequence were derived from a large number of template protein structures; the target sequence was aligned to the backbone of a template structure copying the geometric coordinates of the template to the target sequence. Second, target regions, where geometric constrains could not be copied from the template easily, were modeled. These regions represented either deletions or insertions with respect to the template. The third step involved loop selection and side chain modeling, where a collection of independent models was obtained. Fourth, the final models were scored and ranked, after they had been stereochemically tested and evaluated with a built-in module for protein geometry error checking. Since several models were calculated for the same target, we had to choose an evaluation method that would rank them. As mentioned above the DOPE approach was chosen. It is important to clarify that the DOPE score is not an absolute measure rather a relative one. It has only meaning during the relative comparison of models that have been generated based on the same initial alignment. DOPE method is considered more advanced to its rival GA341 method that is also available by Modeller.

Furthermore, Electrostatic potential surfaces were calculated by solving the nonlinear Poisson–Boltzmann equation using finite difference method as implemented in the PyMOL Software [50]. The potential was calculated on grid points per side (65, 65, 65) and the grid fill by solute parameter was set to 80 %. The dielectric constants of the solvent and the solute were set to 80.0 and 2.0, respectively. An ionic exclusion radius of 2.0 Å, a solvent radius of 1.4 Å, and a solvent ionic strength of 0.145 M were applied. Amber99 [51] charges and atomic radii were used for this calculation.

The models were initially subjected to energy minimization using the Gromacs-implemented force fields to remove the geometrical strain [52]. The model was subsequently solvated with Simple Point Charge (SPC) water using the truncated octahedron box extending to 7 Å from the model. Molecular dynamics were performed after that at 300 K, 1 atm with 2 fs step size, and for a total of ten nanoseconds, using the NVT ensemble in a canonical environment; NVT stands for Number of atoms, Volume, and Temperature that remain constant throughout the simulation. The results of the molecular dynamics simulation were collected into a database by Gromacs which can be subjected to further analysis. The resultant models were initially evaluated within the Gromacs package, version 4.5.5 [53, 54] by a residue packing quality function, which depends on the number of buried nonpolar side chain groups and on hydrogen bonding. Moreover, the suite PROCHECK [55] was employed to further evaluate the quality of the produced TbTOP1B model. Finally, Verify3D [56] was used to evaluate whether the model of TbTOP1B is similar to any known protein structures. *The resultant homology models were visualized using PyMOL* [50] (Fig. 6.1).

Molecular dynamics on the wild-type template and the mutant proteins are already in progress. This will be really helpful in order to observe the flexibility and the kinetics of the protein and notice the shape-size changes, as well as the

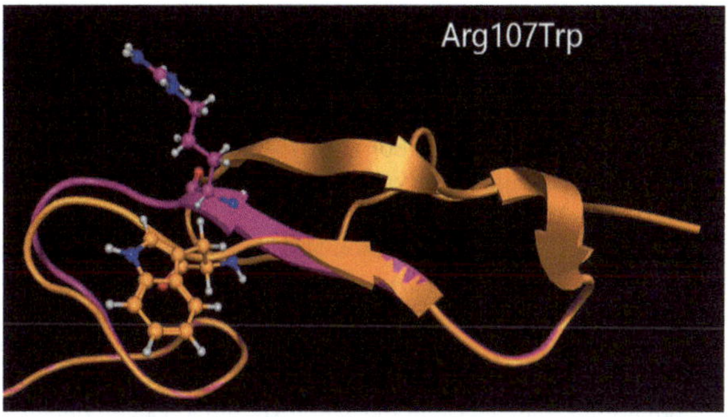

Fig. 6.1 Preliminary work on the R107T point mutation of Notch3 that indicates significant loss of the 3D structural conformation of the b-sheet formation

changes on physicochemical and electrostatic properties. Any changes on conformation and molecular interactions will mean disruption of protein function.

Conclusions

Summarizing, this manuscript focuses on Notch signaling and mutations and simultaneously presenting a 3D Molecular Modeling study on the HUMAN NOTCH EGFS Protein. The first part describes the Notch signaling pathway, citing the contribution of Notch proteins in the embryogenesis, nervous system development and function, in cardiovascular development and in the endocrine development, and focusing on the adult nervous system development. The second part represents the mutations of Notch proteins and their relation with two widespread diseases, the Alzheimer disease and the CADASIL disease. Both of these diseases are affecting the adult brain and causing disability and dementia. Recently, the effort of studying the relation between Notch proteins and degenerative disorders is becoming more and more widespread. Based on the existing assumptions the analysis of the relation between Notch signaling pathway and degenerative diseases could result in promising treatment possibilities.

Acknowledgments This study is funded by the Clinical and system—omics for the identification of the Molecular Determinants of established Chronic Kidney Disease (Project acronym: iMODE-CKD). Grant number: 608332. Seventh Framework Programme FP7.

References

1. Louvi A, Artavanis-Tsakonas S (2006) Notch signalling in vertebrate neural development. Nat Rev Neurosci 7(2):93–102. doi:10.1038/nrn1847
2. Theodosiou A, Arhondakis S, Baumann M, Kossida S (2009) Evolutionary scenarios of Notch proteins. Mol Biol Evol 26(7):1631–1640. doi:10.1093/molbev/msp075

3. Austin J, Kimble J (1987) glp-1 is required in the germ line for regulation of the decision between mitosis and meiosis in C. elegans. Cell 51:589–599. doi:10.1016/0092-8674(87)90128-0
4. Levin M (2005) Left-right asymmetry in embryonic development: a comprehensive review. Mech Dev 122(1):3–25. doi:10.1016/j.mod.2004.08.006
5. Costa RM, Honjo T, Silva AJ (2003) Learning and memory deficits in Notch mutant mice. Curr Biol 13(15):1348–1354
6. Poulson DF (1937) Chromosomal deficiencies and the embryonic development of drosophila melanogaster. Proc Natl Acad Sci U S A 23(3):133–137
7. Redmond L, Oh SR, Hicks C, Weinmaster G, Ghosh A (2000) Nuclear Notch1 signaling and the regulation of dendritic development. Nat Neurosci 3(1):30–40. doi:10.1038/71104
8. Scheer N, Groth A, Hans S, Campos-Ortega JA (2001) An instructive function for Notch in promoting gliogenesis in the zebrafish retina. Development 128(7):1099–1107
9. Kume T (2012) Ligand-dependent Notch signaling in vascular formation. Adv Exp Med Biol 727:210–222. doi:10.1007/978-1-4614-0899-4_16
10. Jensen J (2004) Gene regulatory factors in pancreatic development. Dev Dyn 229(1):176–200. doi:10.1002/dvdy.10460
11. Crosnier C, Vargesson N, Gschmeissner S, Ariza-McNaughton L, Morrison A, Lewis J (2005) Delta-Notch signalling controls commitment to a secretory fate in the zebrafish intestine. Development 132(5):1093–1104. doi:10.1242/dev.01644
12. Yamada T, Yamazaki H, Yamane T, Yoshino M, Okuyama H, Tsuneto M, Kurino T, Hayashi S, Sakano S (2003) Regulation of osteoclast development by Notch signaling directed to osteoclast precursors and through stromal cells. Blood 101:2227–2234. doi:10.1182/blood-2002-06-1740
13. Mohr OL (1919) Character changes caused by mutation of an entire region of a chromosome in drosophila. Genetics 4(3):275–282
14. Morgan TH (1917) Goodale's experiments on gonadectomy of fowls. Science 45(1168):483–484. doi:10.1126/science.45.1168.483
15. Artavanis-Tsakonas S, Simpson P (1991) Choosing a cell fate: a view from the Notch locus. Trends Genet 7(11–12):403–408
16. Greenwald IS, Sternberg PW, Horvitz HR (1983) The lin-12 locus specifies cell fates in Caenorhabditis elegans. Cell 34:435–444
17. Kortschak RD, Tamme R, Lardelli M (2001) Evolutionary analysis of vertebrate Notch genes. Dev Genes Evol 211(7):350–354
18. Maine E, Lissemore J, Starmer W (1995) A phylogenetic analysis of vertebrate and invertebrate notch-related genes. Mol Phylogenet Evol 4:139–149
19. Putnam NH, Butts T, Ferrier DE, Furlong RF, Hellsten U, Kawashima T, Robinson-Rechavi M, Shoguchi E, Terry A, Yu JK, Benito-Gutiérrez EL, Dubchak I, Garcia-Fernàndez J, Gibson-Brown JJ, Grigoriev IV, Horton AC, de Jong PJ, Jurka J, Kapitonov VV, Kohara Y, Kuroki Y, Lindquist E, Lucas S (2008) The amphioxus genome and the evolution of the chordate karyotype. Nature 453:1064–1071
20. Presente A, Andres A, Nye JS (2001) Requirement of Notch in adulthood for neurological function and longevity. Neuroreport 12(15):3321–3325
21. Mathieu P, Martino Adami PV, Morelli L (2013) Notch signaling in the pathologic adult brain. Biomol Concepts 4(5):465–476. doi:10.1515/bmc-2013-0006
22. Alberi L, Hoey SE, Brai E, Scotti AL, Marathe S (2013) Notch signaling in the brain: in good and bad times. Ageing Res Rev 12(3):801–814. doi:10.1016/j.arr.2013.03.004
23. Arumugam TV, Chan SL, Jo DG, Yilmaz G, Tang SC, Cheng A, Gleichmann M, Okun E, Dixit VD, Chigurupati S, Mughal MR, Ouyang X, Miele L, Magnus T, Poosala S, Granger DN, Mattson MP (2006) Gamma secretase-mediated Notch signaling worsens brain damage and functional outcome in ischemic stroke. Nat Med 12(6):621–623. doi:10.1038/nm1403

24. Ferrari Toninelli G, Bernardi C, Quarto M, Lozza G, Memo M, Grilli M (2003) Long-lasting induction of Notch2 in the hippocampus of kainate-treated adult mice. Neuroreport 14(7):917–921. doi:10.1097/01.wnr.0000069962.11849.e6
25. Li L, Krantz I, Deng Y, Genin A, Banta A, Collins C, Qi M, Trask B, Kuo W, Cochran J, Costa T, Pierpont M, Rand EB, Piccoli D, Hood L, Spinner N (1997) Alagille syndrome is caused by mutations in human Jagged1, which encodes a ligand for Notch1. Nat Genet 16 (3):243–251
26. Joutel A, Corpechot C, Ducros A, Vahedi K, Chabriat H, Mouton P, Alamowitch S, Domenga V, Cecillion M, Marechal E, Maciazek J, Vayssiere C, Cruaud C, Cabanis EA, Ruchoux MM, Weissenbach J, Bach JF, Bousser MG, Tournier-Lasserve E (1996) Notch3 mutations in CADASIL, a hereditary adult-onset condition causing stroke and dementia. Nature 383(6602):707–710. doi:10.1038/383707a0
27. Simpson MA, Irving MD, Asilmaz E, Gray MJ, Dafou D, Elmslie FV, Mansour S, Holder SE, Brain CE, Burton BK, Kim KH, Pauli RM, Aftimos S, Stewart H, Kim CA, Holder-Espinasse-M, Robertson SP, Drake WM, Trembath RC (2011) Mutations in NOTCH2 cause Hajdu-Cheney syndrome, a disorder of severe and progressive bone loss. Nat Genet 43(4):303–305. doi:10.1038/ng.779
28. Fischer DF, van Dijk R, Sluijs JA, Nair SM, Racchi M, Levelt CN, van Leeuwen FW, Hol EM (2005) Activation of the Notch pathway in Down syndrome: cross-talk of Notch and APP. FASEB J 19(11):1451–1458. doi:10.1096/fj.04-3395.com
29. Marchesi VT (2014) Alzheimer's disease and CADASIL are heritable, adult-onset dementias that both involve damaged small blood vessels. Cell Mol Life Sci 71(6):949–955
30. Meshorer E, Gruenbaum Y (2008) Gone with the Wnt/Notch: stem cells in laminopathies, progeria, and aging. J Cell Biol 181(1):9–13. doi:10.1083/jcb.200802155
31. Bonini SA, Ferrari-Toninelli G, Montinaro M, Memo M (2013) Notch signalling in adult neurons: a potential target for microtubule stabilization. Ther Adv Neurol Disord 6(6):375–385. doi:10.1177/1756285613490051
32. Guerreiro R, Lohmann E, Kinsella E, Brás J, Luu N, Gurunlian N, Dursun B, Bilgic B, Santana I, Hanagasi H, Gurvit H, Gibbs J, Oliveira C, Emre M, Singleton A (2012) Exome sequencing reveals an unexpected genetic cause of disease: NOTCH3 mutation in a Turkish family with Alzheimer's disease. Neurobiol Aging 33(5):1008.e17–1008.e23. doi:10.1016/j.neurobiolaging.2011.10.009
33. Irvin DK, Zurcher SD, Nguyen T, Weinmaster G, Kornblum HI (2001) Expression patterns of Notch1, Notch2, and Notch3 suggest multiple functional roles for the Notch-DSL signaling system during brain development. J Comp Neurol 436(2):167–181
34. Ethell DW (2010) An amyloid-notch hypothesis for Alzheimer's disease. Neuroscientist 16 (6):614–617. doi:10.1177/1073858410366162
35. Berezovska O, Xia MQ, Hyman BT (1998) Notch is expressed in adult brain, is coexpressed with presenilin-1, and is altered in Alzheimer disease. J Neuropathol Exp Neurol 57:738–745
36. Nagarsheth MH, Viehman A, Lippa SM, Lippa CF (2006) Notch-1 immunoexpression is increased in Alzheimer's and Pick's disease. J Neurol Sci 244:111–116
37. Ishikura N, Clever JL, Bouzamondo-Bernstein E, Samayoa E, Prusiner SB, Huang EJ, DeArmond SJ (2005) Notch-1 activation and dendritic atrophy in prion disease. Proc Natl Acad Sci U S A 102(3):886–891. doi:10.1073/pnas.0408612101
38. Thijs V, Robberecht W, De Vos R, Sciot R (2003) Coexistence of CADASIL and Alzheimer's disease. J Neurol Neurosurg Psychiatry 74(6):790–792
39. Joutel A, Vahedi K, Corpechot C, Troesch A, Chabriat H, Vayssiere C, Cruaud C, Maciazek J, Weissenbach J, Bousser MG, Bach JF, Tournier-Lasserve E (1997) Strong clustering and stereotyped nature of Notch3 mutations in CADASIL patients. Lancet 350(9090):1511–1515. doi:10.1016/S0140-6736(97)08083-5
40. Yamaguchi N, Oyama T, Ito E, Satoh H, Azuma S, Hayashi M, Shimizu K, Honma R, Yanagisawa Y, Nishikawa A, Kawamura M, Imai J, Ohwada S, Tatsuta K, Inoue J, Semba K, Watanabe S (2008) NOTCH3 signaling pathway plays crucial roles in the

proliferation of ErbB2-negative human breast cancer cells. Cancer Res 68(6):1881–1888. doi:10.1158/0008-5472.CAN-07-1597

41. Li X, Zhang X, Leathers R, Makino A, Huang C, Parsa P, Macias J, Yuan JX, Jamieson SW, Thistlethwaite PA (2009) Notch3 signaling promotes the development of pulmonary arterial hypertension. Nat Med 15(11):1289–1297. doi:10.1038/nm.2021
42. Alunni A, Krecsmarik M, Bosco A, Galant S, Pan L, Moens CB, Bally-Cuif L (2013) Notch3 signaling gates cell cycle entry and limits neural stem cell amplification in the adult pallium. Development 140(16):3335–3347. doi:10.1242/dev.095018
43. Ungaro C, Mazzei R, Conforti FL, Sprovieri T, Servillo P, Liguori M, Citrigno L, Gabriele AL, Magariello A, Patitucci A, Muglia M, Quattrone A (2009) CADASIL: extended polymorphisms and mutational analysis of the NOTCH3 gene. J Neurosci Res 87(5):1162–1167. doi:10.1002/jnr.21935
44. Schmidt H, Zeginigg M, Wiltgen M, Freudenberger P, Petrovic K, Cavalieri M, Gider P, Enzinger C, Fornage M, Debette S, Rotter JI, Ikram MA, Launer LJ, Schmidt R (2011) Genetic variants of the NOTCH3 gene in the elderly and magnetic resonance imaging correlates of age-related cerebral small vessel disease. Brain 134(Pt 11):3384–3397. doi:10.1093/brain/awr252
45. Guruharsha KG, Kankel MW, Artavanis-Tsakonas S (2012) The Notch signalling system: recent insights into the complexity of a conserved pathway. Nat Rev Genet 13(9):654–666
46. Artavanis-Tsakonas S, Muskavitch MA (2010) Notch: the past, the present, and the future. Curr Top Dev Biol 92:1–29. doi:10.1016/S0070-2153(10)92001-2
47. Fre S, Huyghe M, Mourikis P, Robine S, Louvard D, Artavanis-Tsakonas S (2005) Notch signals control the fate of immature progenitor cells in the intestine. Nature 435(7044):964–968. doi:10.1038/nature03589
48. Sali A, Potterton L, Yuan F, van Vlijmen H, Karplus M (1995) Evaluation of comparative protein modeling by MODELLER. Proteins 23(3):318–326. doi:10.1002/prot.340230306
49. Cordle J, Johnson S, Tay J, Roversi P, Wilkin M, de Madrid B, Shimizu H, Jensen S, Whiteman P, Jin B, Redfield C, Baron M, Lea S, Handford P (2008) A conserved face of the Jagged/Serrate DSL domain is involved in Notch trans-activation and cis-inhibition. Nat Struct Mol Biol 15(8):849–857. doi:10.1038/nsmb.1457
50. DeLano W (2002) The PyMOL user's manual San Carlos. DeLano Scientific, Palo Alto, CA, USA
51. Duan Y, Wu C, Chowdhury S, Lee MC, Xiong G, Zhang W, Yang R, Cieplak P, Luo R, Lee T, Caldwell J, Wang J, Kollman P (2003) A point-charge force field for molecular mechanics simulations of proteins based on condensed-phase quantum mechanical calculations. J Comput Chem 24(16):1999–2012. doi:10.1002/jcc.10349
52. Hess B, Kutzner C, van der Spoel D, Lindahl E (2008) GROMACS 4: algorithms for highly efficient, load-balanced, and scalable molecular simulation. J Chem Theory Comput 4:435–447
53. Hess B, Kutzner C, van der Spoel D, Lindahl E (2008) GROMACS 4: algorithms for highly efficient, load-balanced, and scalable molecular simulation. J Chem Theor Comput 4(3):435–447. doi:10.1021/ct700301q
54. Van Der Spoel D, Lindahl E, Hess B, Groenhof G, Mark AE, Berendsen HJ (2005) GROMACS: fast, flexible, and free. J Comput Chem 26(16):1701–1718. doi:10.1002/jcc.20291
55. Laskowski RA, Rullmannn JA, MacArthur MW, Kaptein R, Thornton JM (1996) AQUA and PROCHECK-NMR: programs for checking the quality of protein structures solved by NMR. J Biomol NMR 8(4):477–486
56. Eisenberg D, Luthy R, Bowie JU (1997) VERIFY3D: assessment of protein models with three-dimensional profiles. Methods Enzymol 277:396–404

Chapter 7
Pro-oxidant DNA Breakage Induced by the Interaction of L-DOPA with Cu(II): A Putative Mechanism of Neurotoxicity

Asma Perveen, Husain Yar Khan, S.M. Hadi, Ghazi A. Damanhouri, Ahmed Alharrasi, Shams Tabrez, and Ghulam Md Ashraf

Abstract There are reports in scientific literature that the concentration of copper ions in Parkinsonian brain is at a level that could promote oxidative DNA damage. The possibility of copper chelation by antioxidants excited us to explore the generation of reactive oxygen species (ROS) and DNA damage by the interaction of L-DOPA with Cu(II) ions. In the present manuscript, L-DOPA was tested for its ability to bind with Cu(II) and reduce it to Cu(I). The generation of ROS, such as superoxide anion (O_2^-) and hydroxyl radical ($OH^\bullet$), was also ascertained. As a result of L-DOPA and Cu(II) interaction, the generation of O_2^- was found to be increased in a time-dependent manner. Moreover, the formation of $OH^\bullet$ was also found to be enhanced with increasing concentrations of L-DOPA. Furthermore, Comet assay results clearly showed significantly higher cellular DNA breakage in lymphocytes treated with L-DOPA and Cu(II) as compared to those that were treated with L-DOPA alone. However, such DNA degradation was inhibited to a significant extent by scavengers of ROS and neocuproine, a membrane permeable Cu(I)-specific sequestering agent. These findings demonstrate that L-DOPA exhibits a pro-oxidant activity in the presence of copper ions.

A. Perveen • S.M. Hadi
Department of Biochemistry, Faculty of Life Sciences, Aligarh Muslim University, Aligarh, Uttar Pradesh 202002, India

H.Y. Khan
Department of Biochemistry, Faculty of Life Sciences, Aligarh Muslim University, Aligarh, Uttar Pradesh 202002, India

Chair of Oman's Medicinal Plants and Marine Natural Products, University of Nizwa, Birkat Al-mouz, PC 616, Nizwa, Sultanate of Oman

G.A. Damanhouri • S. Tabrez (✉) • G.M. Ashraf (✉)
King Fahd Medical Research Center, King Abdulaziz University, P.O. Box 80216, Jeddah 21589, Saudi Arabia
e-mail: shamstabrez1@gmail.com; ashraf.gm@gmail.com; gashraf@kau.edu.sa

A. Alharrasi
Chair of Oman's Medicinal Plants and Marine Natural Products, University of Nizwa, Birkat Al-mouz, PC 616, Nizwa, Sultanate of Oman

P. Vlamos, A. Alexiou (eds.), *GeNeDis 2014*, Advances in Experimental Medicine and Biology 822, DOI 10.1007/978-3-319-08927-0_7

Keywords L-DOPA • DNA breakage • Copper • Comet assay • ROS • Parkinson's disease

7.1 Introduction

L-3,4-Dihydroxy phenylalanine (L-DOPA) is an important metabolite in various metabolic reactions with a chemical formula of $C_9H_{11}NO_4$, and molar mass of 197.19 g/mol [1]. Dopamine is a neurotransmitter in the central nervous system (CNS) and accounts for 90 % of the total catecholamines. It is formed by the decarboxylation of L-DOPA which in turn is formed by hydroxylation of tyrosine [2, 3].

Parkinson's disease (PD) is a progressive and degenerative movement disorder and results from degeneration of dopamine-releasing neurons of the substantia nigra. The neurological disorder in PD is associated with an underproduction of dopamine in the human brain [2, 3]. L-DOPA could cross the protective blood-brain barrier (BBB), whereas in the form of dopamine it cannot cross BBB. In view of the earlier mentioned fact, L-DOPA has been suggested as an effective drug in the treatment of PD [4]. Once L-DOPA entered to the CNS, it is converted into dopamine by the enzyme aromatic L-amino acid decarboxylase [2]. Pyridoxal phosphate (vitamin B6) is a cofactor required for this reaction, and occasionally administered along with L-DOPA.

It has been reported that Cu_2SO_4 pretreatment could block lipid peroxidation (LPO) and prevent the striatal dopamine depletion induced by the administration of 1-methyl-4-phenylpiridinium (MPP+), the toxic metabolite of 1-methyl-4-phenyl-1,2,3,6-tetrahydropyridine (MPTP), in a Parkinson's mice model [5]. In one study, Zubeldia et al. [6] showed that the preservation of tyrosine hydroxylase activity participates in the neuroprotective effects derived from the copper supplementation in an experimental Parkinson's model. However, copper has also been reported to be neurotoxic as evident by the brain pathology of patients with copper overload as a result of Wilson's disease [7]. Some studies suggest that LPO is promoted by copper ions [8, 9], which also catalyzes the formation of highly reactive hydroxyl radicals ($OH^{\bullet}$) from hydrogen peroxide [10]. Copper ions and hydrogen peroxide could lead to DNA damage, strand breaks [11], and chemical changes in purine and pyrimidine bases; especially conversion of guanine into 8-hydroxyguanine [12]. In one study, Spencer et al. [13] has shown that L-DOPA, dopamine, and 3-O-methyl-DOPA could cause extensive base modification in DNA in the presence of hydrogen peroxide and traces of copper ions [13]. They suggested that release of copper ion, in the presence of L-DOPA and its metabolites, may be an important mechanism of neurotoxicity involved in PD and amyotrophic lateral sclerosis (ALS).

A number of studies from our laboratory have shown that many known antioxidants such as plant polyphenols [14, 15], ascorbic acid [16], uric acid [17], and thymoquinone [18] can generate ROS in the presence of transition metal ions such as copper results into oxidative DNA breakage. In the light of our studies in the past two decades, we have suggested that several of the biological antioxidants are themselves capable of generating ROS and may act as pro-oxidants under appropriate conditions such as in the presence of copper [16, 19, 20]. In view of this and the fact that L-DOPA is produced from tyrosine by a copper-containing enzyme, tyrosine hydroxylase and presents a possibility of copper chelation, we considered it quite worthwhile to explore the generation of active oxygen species and breakage to cellular DNA by the L-DOPA-Cu(II) system.

It has been proposed that considerable DNA damage may be caused by endogenous metabolites produced during the body's normal metabolic processes. In an earlier study from our laboratory, it was demonstrated that L-DOPA is capable of binding to DNA [21]. DNA damage induced by dopamine in the presence of copper ions has been also reported by Li and Cao [22]. In this study, we have shown that L-DOPA can also bind to Cu(II) and has the ability to reduce it. Furthermore, using a cellular system of lymphocytes isolated from human peripheral blood and alkaline single cell gel electrophoresis (Comet assay), we have confirmed that L-DOPA in the presence of Cu(II) ions is capable of causing DNA degradation in cells such as lymphocytes. It is also shown that ROS are involved in this DNA breakage reaction.

7.2 Materials and Methods

7.2.1 Spectrophotometric Study of the Reduction of Cu (II) to Cu (I) by L-DOPA

Selective copper sequestering agent, neocuproine, was employed to detect reduction of Cu(II) to Cu(I) spectrophotometrically by recording the spectra between 300 and 550 nm. The reaction mixture (3.0 ml) contained 10 mM Tris–HCl (pH 7.5), 50 μM L-DOPA, 50 μM Cu(II), and 100 μM neocuproine. The reactions were started by the addition of neocuproine and the spectra were recorded immediately.

7.2.2 Stoichiometric Titration of Cu(I) Production

The stoichiometry of the L-DOPA-Cu(II) interaction was determined by titration with neocuproine. L-DOPA (50 μM) in 10 mM Tris–HCl (pH 7.5) was mixed with varying concentrations of Cu(II) and 100 μM neocuproine in a total reaction volume of 3.0 ml. Absorbance was recorded at 450 nm after 10 min of incubation at room temperature.

7.2.3 Detection of Superoxide Anion (O_2^-) Generation by L-DOPA

Superoxide anion production by L-DOPA in the presence of 50 μM Cu(II) was detected by the reduction of nitrobluetetrazolium (NBT) essentially as described by Nakayama et al. [23]. A typical assay mixture contains 50 mM sodium phosphate buffer pH 8.0, 33.0 μM NBT, 0.1 mM EDTA, and 0.06 % Triton X-100 in a total volume of 3.0 ml. The reaction was started by the addition of L-DOPA. Immediately after mixing, the absorbance was recorded at 560 nm at different time intervals against a blank that did not contain L-DOPA.

7.2.4 Assay for Hydroxyl Radical (OH^-) Generation by L-DOPA and Cu(II)

In order to determine the hydroxyl radical production by increasing concentrations of L-DOPA in the presence of 50 μM Cu(II), the method of Quinlan and Gutteridge was followed [24]. Calf thymus DNA (100 μg) was used as a substrate and the malondialdehyde generated from deoxyribose radicals was assayed by recording the absorbance at 532 nm.

7.2.5 Isolation of Lymphocytes

Fresh heparinized blood samples (2.0 ml) from a single nonsmoking healthy donor were obtained by venipuncture and diluted suitably in Ca^{++}- and Mg^{++}-free PBS for each experiment. Lymphocytes were isolated from blood using Histopaque 1077 (Sigma Diagnostics, St. Louis, MO, USA), and the cells were finally suspended in RPMI 1640.

7.2.6 Viability Assessment of Lymphocytes

The lymphocytes were checked for their viability before the start and after the end of the reaction using Trypan blue exclusion test [25]. The viability of the cells was found to be greater than 93 %.

7.2.7 *Treatment of Lymphocytes*

Lymphocytes (1×10^5 cells) were exposed to different concentrations of L-DOPA in the absence and presence of $CuCl_2$ at a concentration of 50 μM in a total reaction volume of 1 ml (400 μl RPMI, PBS Ca^{2+} and Mg^{2+} free, and indicated concentrations of L-DOPA and $CuCl_2$). Incubation was performed at 37 °C for 1 h. In another set of experiments, scavengers of ROS or neocuproine (copper chelator) were added at the final concentrations indicated. After the incubation, the reaction mixture was centrifuged at 4,000 rpm, the supernatant was discarded, and pelleted lymphocytes were resuspended in 100 μl of PBS and processed further for Comet assay.

7.2.8 *Estimation of DNA Breakage by Comet Assay*

Comet assay was performed under alkaline conditions essentially according to the procedure of Singh et al. [26] with slight modifications as described by us in detail previously [27].

7.2.9 *Statistical Analysis*

The statistical analysis was performed as described by Tice et al. [28] and is expressed as mean ± S.E.M. of three independent experiments. A Student's *t*-test was used to examine statistically significant differences. Analysis of variance was performed using ANOVA. P-values <0.05 were considered statistically significant.

7.3 Results

7.3.1 *Reduction of Cu(II) to Cu(I) by L-DOPA*

The production of Cu(I), as a result of reduction of Cu(II) by L-DOPA, was analyzed using neocuproine, a Cu(I)-sequestering agent that binds specifically to the reduced form of copper, Cu(I), but not to the oxidized form, Cu(II). The Cu(I)-neocuproine complex absorbs maximally at 450 nm. As shown in Fig. 7.1, neither Cu(II) nor L-DOPA interfered with the maxima, whereas L-DOPA + Cu(II) reacted to generate Cu(I) which complexed with neocuproine to generate a peak at 450 nm. The results show that L-DOPA is able to reduce Cu(II) to Cu(I) contributing to its redox cycling.

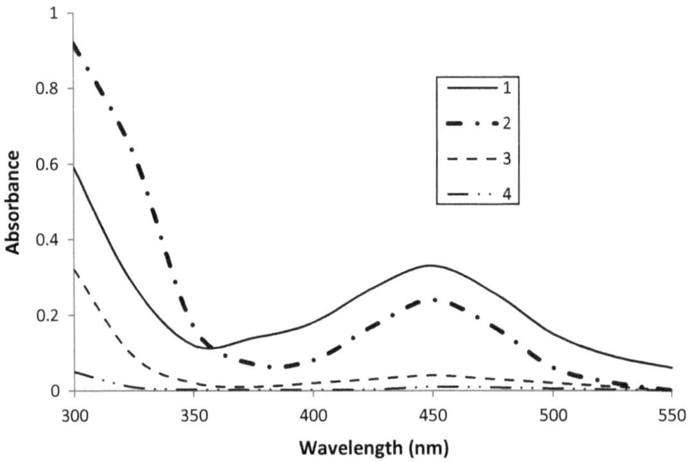

Fig. 7.1 Detection of L-DOPA-induced Cu(I) production. Trace 1: Neocuproine + 50 μM L-DOPA + 50 μM Cu(II); Trace 2: Neocuproine + 50 μM Cu(I); Trace 3: Neocuproine + 50 μM Cu(II); Trace 4: Neocuproine + 50 μM L-DOPA. In all cases the reaction was started by the addition of neocuproine (100 μM)

7.3.2 *Stoichiometry of L-DOPA and Cu(II) Interaction*

To determine the stoichiometry of Cu(II) reduction by L-DOPA, we again used the Cu(I) sequestering agent, namely neocuproine. A Job plot of absorbance versus [Cu(II)/L-DOPA] (Fig. 7.2) showed maximum absorbance at a ratio of 3. This implies a 3:1 stoichiometry for the reduction of Cu(II) by L-DOPA. In other words, every molecule of L-DOPA can bind with 3 copper ions.

7.3.3 *Generation of Superoxide Anions (O_2^-) by L-DOPA in the Presence of Cu(II) with Increasing Incubation Time*

The production of superoxide anions was determined by the method of Nakayama et al. [23], which involves reduction of NBT by L-DOPA-Cu(II) to a formazan. The time-dependent generation of superoxide anions by L-DOPA and Cu(II), as evidenced by the increase in absorbance at 560 nm, is shown in Fig. 7.3. The fact that NBT was genuinely assaying superoxide was confirmed by superoxide dismutase (SOD) (100 μg/ml) inhibiting the reaction (results not shown). It is known that superoxide undergoes automatic dismutation at neutral pH to form H_2O_2 which in the presence of transition metals such as copper favors Fenton-type reaction to generate hydroxyl radicals which could act as a proximal DNA cleaving agent leading to oxidative DNA breakage.

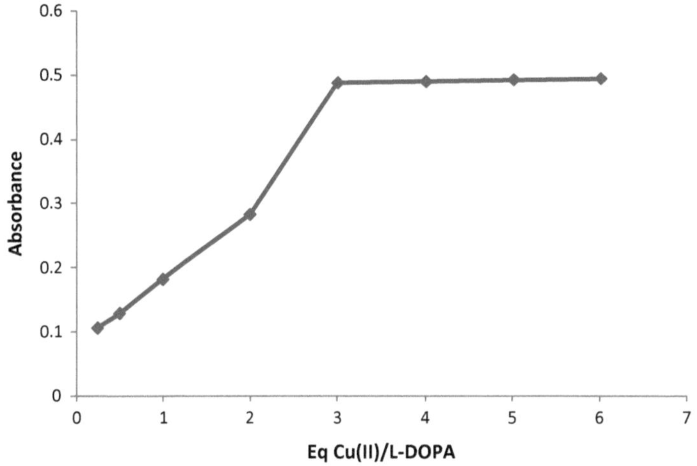

Fig. 7.2 Stoichiometry of L-DOPA and Cu(II) interaction. The concentration of L-DOPA was 50 μM in the presence of 100 μM neocuproine. The absorbance of samples at 450 nm is plotted against equivalents of Cu(II) per molar equivalent of L-DOPA. All points represent triplicate samples and their mean values are plotted

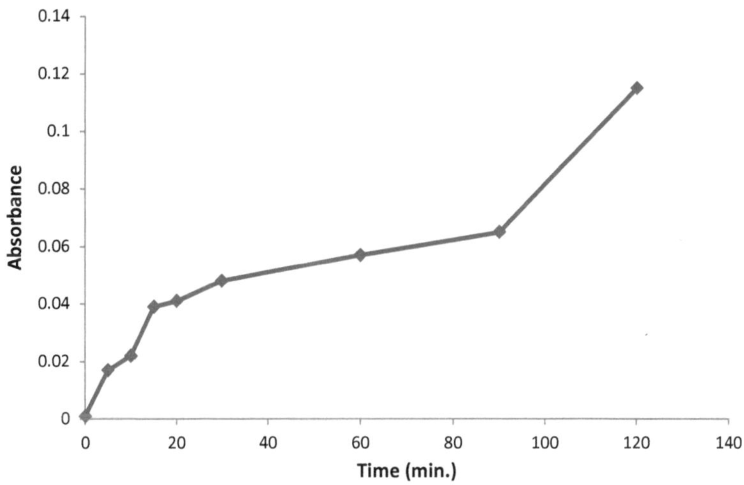

Fig. 7.3 Photogeneration of superoxide anions (O_2^{-}) by L-DOPA in the presence of Cu(II) with increasing incubation time. Reaction mixture contained 50 mM phosphate buffer (pH 8) and 50 μM of L-DOPA and $CuCl_2$ each. The samples were placed at a distance of 10 cm from the light source. All values reported are mean of three independent experiments

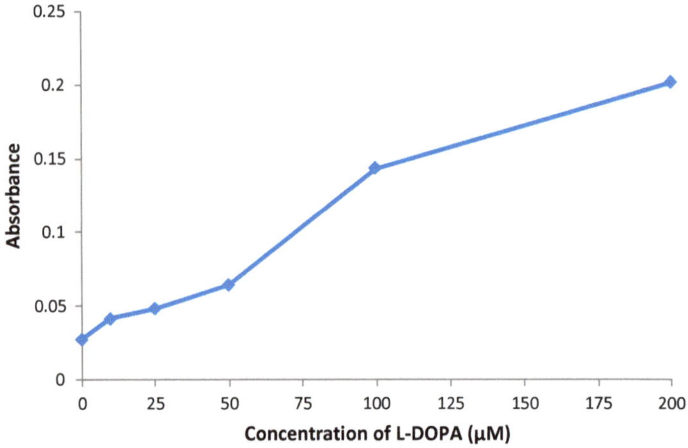

Fig. 7.4 Formation of hydroxyl radicals (OH•) with increasing concentrations of L-DOPA in the presence of Cu(II). Reaction mixture (0.5 ml) contained 100 μg calf thymus DNA as substrate, 50 μM $CuCl_2$, and indicated concentrations of L-DOPA. The reaction mixture was incubated at 37 °C for 30 min. Hydroxyl radical formation was measured by determining the TBA reactive material as described earlier. All values reported are mean ± SEM of three independent experiments

7.3.4 Formation of Hydroxyl Radicals (OH•) with Increasing Concentrations of L-DOPA in the Presence of Cu(II)

It has been previously shown that during the reduction of Cu(II) to Cu(I), hydroxyl radicals are formed which serve as the proximal DNA cleaving agent [29]. Therefore, the ability of L-DOPA to generate hydroxyl radicals in the presence of Cu(II) was examined. The assay is based upon the fact that degradation of DNA by hydroxyl radicals results in the release of 2-thiobarbituric acid (TBA) reactive material, which forms a colored adduct with TBA whose absorbance was read at 532 nm. The results given in Fig. 7.4 clearly show that increasing concentrations of L-DOPA lead to a progressive increase in the formation of hydroxyl radicals.

7.3.5 DNA Breakage by L-DOPA-Cu(II) System in Lymphocytes as Measured by Comet Assay

Increasing concentrations of L-DOPA (0–100 μM) alone or in the presence of 50 μM $CuCl_2$ were tested for DNA breakage in the intact isolated human peripheral lymphocytes using alkaline single cell gel electrophoresis (Comet assay) (Fig. 7.5). The corresponding Comet tail length is plotted as a function of L-DOPA concentration. It was observed that L-DOPA alone causes breakage of cellular DNA; the degree of such DNA breakage was significantly enhanced in the

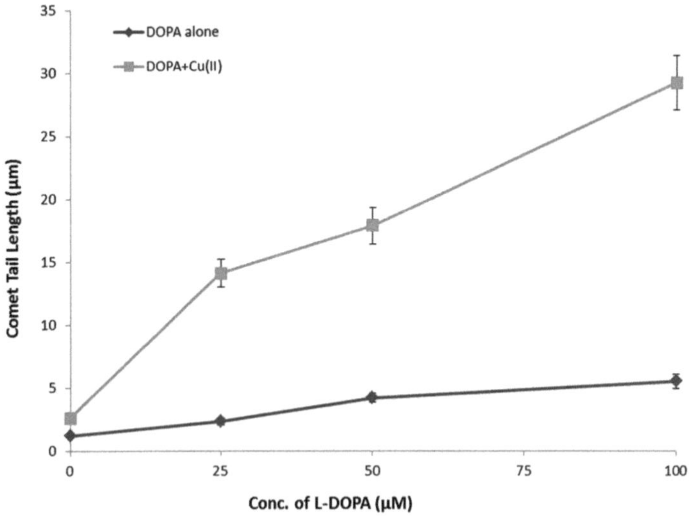

Fig. 7.5 DNA breakage by L-DOPA in human peripheral lymphocytes in the absence and presence of Cu(II). 1×10^5 lymphocytes were incubated in microfuge tubes with reaction mixture at 4 °C for 1 h. Reaction mixture contained RPMI (400 μl), Ca^{2+} and Mg^{2+} free PBS, increasing concentrations of L-DOPA (0–100 μM), alone (*filled diamond*) and with fixed concentration of Cu (II) (50 μM) (*filled square*) and processed further for Comet assay as described earlier. All points represent mean ± SEM of three independent experiments

presence of Cu(II). The result of treatment with Cu(II) (50 μM) alone was similar to untreated lymphocyte without any significant DNA breakage (not shown). The results clearly establish that L-DOPA-Cu(II) system is capable of DNA breakage in isolated lymphocytes. Thus, such cellular DNA breakage is physiologically feasible and could be of biological significance.

7.3.6 *Effect of Copper Chelating Agent Neocuproine on L-DOPA and Cu(II) Induced DNA Breakage in Human Lymphocytes*

Copper chelator neocuproine was used to study its effect on DNA breakage induced by L-DOPA-Cu(II) system in lymphocytes (Fig. 7.6). A clear inhibition was seen in the presence of neocuproine, which is a cell membrane permeable Cu(I) specific chelator, on L-DOPA-Cu(II) induced DNA breakage. These results indicate that Cu(I) is an intermediate in the pathway that leads to DNA breakage induced by L-DOPA and Cu(II).

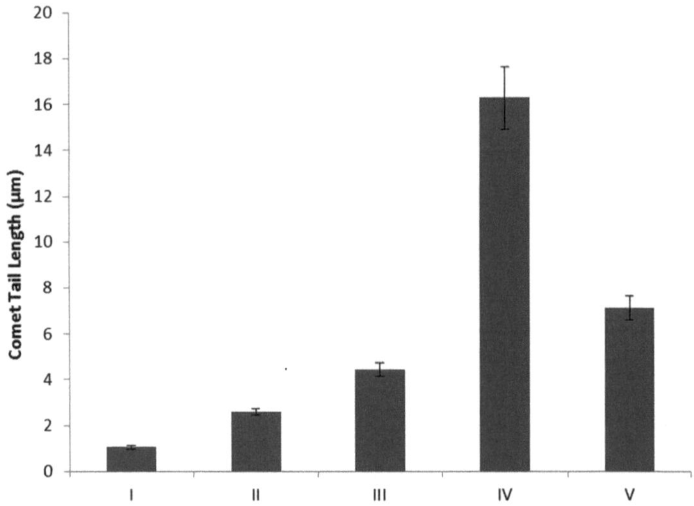

Fig. 7.6 Effect of copper chelating agent neocuproine on L-DOPA + Cu(II) induced DNA breakage in human lymphocytes. I: Untreated control, II: Cu(II) (50 μM), III: L-DOPA (50 μM), IV: L-DOPA (50 μM) + Cu(II) (50 μM), V: L-DOPA (50 μM) + Cu(II) (50 μM) + Neocuproine (100 μM)

7.3.7 Effect of ROS Scavengers on Cellular DNA Breakage Induced by L-DOPA and Cu(II)

Table 7.1 gives the results of an experiment where various scavengers of reactive oxygen species such as superoxide dismutase (SOD), catalase, and thiourea were tested for their effect on L-DOPA-Cu(II) induced DNA breakage in lymphocytes. It is well known that SOD and catalase scavenge superoxide and H_2O_2, respectively and thiourea removes hydroxyl radicals. All scavengers caused a significant inhibition of DNA breakage as evidenced by the decreased Comet tail lengths. It may be mentioned here that due to the site-specific nature of the reaction of hydroxyl radicals with DNA, it is difficult for any trapping molecules to intercept them completely [30]. This possibly accounts for the incomplete inhibition of DNA breakage by the scavengers. The results indicate that superoxide anion and H_2O_2 are essential components in the pathway that leads to the formation of hydroxyl radical and other species which would be the proximal DNA cleaving agents.

7.4 Discussion

Parkinson's disease (PD) affects around 1 % of the population over the age of 60 years in the western countries and similar trends are expected in developing countries due to an increase in life expectancies [2, 3]. The main pathological

Table 7.1 Effect of ROS scavengers on cellular DNA breakage induced by L-DOPA and Cu(II)

Treatment	Comet tail length (μm)	% Inhibition
Untreated (control)	1.06	–
L-DOPA (50 μM) + $CuCl_2$ (50 μM)	16.28	–
+Catalase (100 μg/ml)	7.47	57.88
+SOD (100 μg/ml)	5.26	72.40
+Thiourea (1 mM)	8.54	50.85

Lymphocytes embedded in agarose were layered with reaction mixture containing L-DOPA and $CuCl_2$ in presence of different scavengers of ROS at concentrations indicated at 4 °C for 1 h and further processed for comet assay as described in Sect. 2. Values reported are mean ± SEM of three independent experiments

feature of PD is the progressive destruction of dopamine-producing cells in the substantia nigra region of the brain stem. The loss of dopamine production affects the balance between dopamine and acetylcholine in the brain [2, 3]. The actual reason behind the destruction of dopamine-producing cells has puzzled researchers for several years but a consensus is now emerging that oxidative stress and metal toxicity is the leading cause of PD [3]. Numerous studies have shown that PD patients have low levels of natural antioxidants (glutathione and superoxide dismutase) and high levels of metals such as iron and copper in the substantia nigra area of their brains [31].

Copper is one of the several metal ions that are required for essential body functions but is toxic in excess. Copper is present throughout the brain and is most prominent in the basal ganglia, hippocampus, cerebellum, numerous synaptic membranes, and in the cell bodies of cortical pyramidal and cerebellar granular neurons [32, 33]. Enzymes in the central nervous system that depend on copper for their function include tyrosinase, copper/zinc superoxide dismutase, ceruloplasmin, hephaestin, and cytochrome c oxidase [34]. Copper is also implicated directly or indirectly in the pathogenesis of numerous neurological diseases, including Alzheimer's disease, Amyotrophic lateral sclerosis, Huntington's disease, PD, Prion disease, and Wilson's disease [31, 32].

In one study, Pall et al. [35] reported significantly higher copper concentration in cerebrospinal fluid of 24 idiopathic PD patients compared with the control. In view of their study, they suggested possible role of copper-catalyzed oxidative mechanisms in the pathogenesis of PD. As mentioned above, there could be an excess accumulation of copper in PD brain. Copper ions can therefore interact with endogenous metabolites such as L-DOPA in the presence of molecular oxygen to generate ROS by Haber-Weiss and/or Fenton reactions, eventually leading to oxidative damage to biomolecules [10]. The pro-oxidant nature of flavonoids isolated from *Rhus javanica* in the presence of Cu(II) was highlighted by Lin et al. [36]. They reported breakage of supercoiled plasmid pBR322 DNA in the presence of Cu(II) via different ROS species by various isolated flavonoids. Recently, a concentration-dependent pBR322 plasmid DNA strand breakage as a result of Cu(II)-dependent pro-oxidant action of knipholone anthrone has been also reported by Habtemariam and Dagne [37]. In the current study we examined the precise role of copper in such a reaction mechanism and provided the evidence for cellular DNA degradation by the interaction of copper with L-DOPA via ROS generation.

Here we have shown that L-DOPA in the presence of Cu(II) is capable of causing strand scission in DNA and that this reaction is associated with generation of ROS which act as proximal DNA cleaving agents. It is generally believed that the reaction of hydroxyl radicals with DNA is preceded by the association of a complex with DNA, followed by in situ generation of hydroxyl radicals [38]. Therefore in the presence of copper ions, L-DOPA, which is capable of binding to DNA, may form a ternary complex of L-DOPA-Cu(II)-DNA as is the case with several other ROS generating and DNA breakage systems [20, 29]. Presumably in this ternary complex L-DOPA would reduce Cu(II) to Cu(I), leading to the generation of hydroxyl radicals by Fenton reaction. These hydroxyl radicals would ultimately cause oxidative cleavage of cellular DNA. Therefore, this study explains the mechanism of cellular DNA breakage by L-DOPA in the presence of Cu(II). It further emphasizes the putative role of L-DOPA and copper ions as a source of oxidative DNA damage and neurotoxicity in vivo.

Conclusion

Dopaminergic cell death associated with PD is largely linked to oxidative stress aggravation by metal toxicity and recent studies also suggested the importance of thiol-redox signaling regulating neuronal loss [39]. The data presented in this article provides a mechanistic insight to explain the pathogenesis of the disease in the light of excess copper and its interaction with endogenous metabolites. The use of antioxidants to supplement the treatment of patients suffering different pathologies that are associated to iron and/or copper overload and oxidative stress, such as Parkinson's and Alzheimer's diseases, is widely accepted. However, our data highlights the importance of considering metal binding when identifying antioxidants to treat and prevent neurodegenerative disorders [40, 41]. In view of our results, we believe, thiol-based antioxidants rather than polyphenol-based antioxidants could fulfill both the requirements. Based on the scientific reports as well as our results, further research is required to elucidate novel therapeutic approaches against oxidative stress and redox imbalance in PD as well.

Acknowledgements The authors gratefully acknowledge the research facilities provided by Aligarh Muslim University, Aligarh, India. Thanks are also due to King Fahd Medical Research Center, King Abdulaziz University (Jeddah, Saudi Arabia) for other facilities.

Conflict of interest: Declared none.

References

1. Sergutina AV, Rakhmanova VI (2013) The effects of L-DOPA on cytochemical indices of oxidative metabolism in the brains of rats with different levels of motor activity. Neurochem J 7(4):291–295. doi:10.1134/s1819712413030124
2. Tabrez S, Jabir NR, Shakil S, Greig NH, Alam Q, Abuzenadah AM, Damanhouri GA, Kamal MA (2012) A synopsis on the role of tyrosine hydroxylase in Parkinson's disease. CNS Neurol Disord Drug Targets 11(4):395–409
3. Khan MS, Tabrez S, Priyadarshini M, Priyamvada S, Khan MM (2012) Targeting Parkinson's – tyrosine hydroxylase and oxidative stress as points of interventions. CNS Neurol Disord Drug Targets 11(4):369–380
4. Cotzias GC (1968) L-Dopa for Parkinsonism. N Engl J Med 278(11)
5. Alcaraz-Zubeldia M, Rojas P, Boll C, Rios C (2001) Neuroprotective effect of acute and chronic administration of copper (II) sulfate against MPP + neurotoxicity in mice. Neurochem Res 26(1):59–64
6. Alcaraz-Zubeldia M, Boll-Woehrlen MC, Montes-López S, Pérez-Severiano F, Martínez-Lazcano JC, Díaz-Ruiz A, Ríos C (2009) Copper sulfate prevents tyrosine hydroxylase reduced activity and motor deficits in a Parkinson's disease model in mice. Rev Invest Clin 61(5):405–411
7. Hartard C, Weisner B, Dieu C, Kunze K (1993) Wilson's disease with cerebral manifestation: monitoring therapy by CSF copper concentration. J Neurol 241(2):101–107
8. Hochstein P, Kumar KS, Forman SJ (1980) Lipid peroxidation and the cytotoxicity of copper. Ann New York Acad Sci 355:240–248
9. Esterbauer H, Striegl G, Puhl H, Rotheneder M (1989) Continuous monitoring of in vitro oxidation of human low density lipoprotein. Free Radic Res Commun 6(1):67–75
10. Halliwell B, Gutteridge JM (1990) Role of free radicals and catalytic metal ions in human disease: an overview. Methods Enzymol 186:1–85
11. Tachon P (1990) DNA single strand breakage by H_2O_2 and ferric or cupric ions: its modulation by histidine. Free Radic Res Commun 9(1):39–47
12. Aruoma OI, Halliwell B, Gajewski E, Dizdaroglu M (1991) Copper-ion-dependent damage to the bases in DNA in the presence of hydrogen peroxide. Biochem J 273(Pt 3):601–604
13. Spencer JP, Jenner A, Aruoma OI, Evans PJ, Kaur H, Dexter DT, Jenner P, Lees AJ, Marsden DC, Halliwell B (1994) Intense oxidative DNA damage promoted by L-dopa and its metabolites. Implications for neurodegenerative disease. FEBS Lett 353(3):246–250
14. Shamim U, Hanif S, Ullah MF, Azmi AS, Bhat SH, Hadi SM (2008) Plant polyphenols mobilize nuclear copper in human peripheral lymphocytes leading to oxidatively generated DNA breakage: implications for an anticancer mechanism. Free Radic Res 42(8):764–772. doi:10.1080/10715760802302251
15. Khan HY, Zubair H, Ullah MF, Ahmad A, Hadi SM (2012) A prooxidant mechanism for the anticancer and chemopreventive properties of plant polyphenols. Curr Drug Targets 13 (14):1738–1749
16. Ullah MF, Khan HY, Zubair H, Shamim U, Hadi SM (2011) The antioxidant ascorbic acid mobilizes nuclear copper leading to a prooxidant breakage of cellular DNA: implications for chemotherapeutic action against cancer. Cancer Chemother Pharmacol 67(1):103–110. doi:10.1007/s00280-010-1290-4
17. Shamsi FA, Hadi SM (1995) Photoinduction of strand scission in DNA by uric acid and Cu(II). Free Radic Biol Med 19(2):189–196
18. Zubair H, Khan HY, Sohail A, Azim S, Ullah MF, Ahmad A, Sarkar FH, Hadi SM (2013) Redox cycling of endogenous copper by thymoquinone leads to ROS-mediated DNA breakage and consequent cell death: putative anticancer mechanism of antioxidants. Cell Death Dis 4:e660. doi:10.1038/cddis.2013.172
19. Ullah MF, Ahmad A, Khan HY, Zubair H, Sarkar FH, Hadi SM (2013) The prooxidant action of dietary antioxidants leading to cellular DNA breakage and anticancer effects: implications

for chemotherapeutic action against cancer. Cell Biochem Biophys 67(2):431–438. doi:10.1007/s12013-011-9303-4
20. Hadi SM, Bhat SH, Azmi AS, Hanif S, Shamim U, Ullah MF (2007) Oxidative breakage of cellular DNA by plant polyphenols: a putative mechanism for anticancer properties. Semin Cancer Biol 17(5):370–376. doi:10.1016/j.semcancer.2007.04.002
21. Husain S, Hadi SM (1995) Strand scission in DNA induced by L-DOPA in the presence of Cu (II). FEBS Lett 364(1):75–78
22. Li Y, Cao Z (2002) The neuroprotectant ebselen inhibits oxidative DNA damage induced by dopamine in the presence of copper ions. Neurosci Lett 330(1):69–73
23. Nakayama T, Kimura T, Kodama M, Nagata C (1983) Generation of hydrogen peroxide and superoxide anion from active metabolites of naphthylamines and aminoazo dyes: its possible role in carcinogenesis. Carcinogenesis 4(6):765–769
24. Quinlan GJ, Gutteridge JM (1987) Oxygen radical damage to DNA by rifamycin SV and copper ions. Biochem Pharmacol 36(21):3629–3633
25. Pool-Zobel BL, Guigas C, Klein R, Neudecker C, Renner HW, Schmezer P (1993) Assessment of genotoxic effects by lindane. Food Chem Toxicol 31(4):271–283
26. Singh NP, McCoy MT, Tice RR, Schneider EL (1988) A simple technique for quantitation of low levels of DNA damage in individual cells. Exp Cell Res 175(1):184–191
27. Khan HY, Zubair H, Ullah MF, Ahmad A, Hadi SM (2011) Oral administration of copper to rats leads to increased lymphocyte cellular DNA degradation by dietary polyphenols: implications for a cancer preventive mechanism. Biometals 24(6):1169–1178. doi:10.1007/s10534-011-9475-9
28. Tice RR, Agurell E, Anderson D, Burlinson B, Hartmann A, Kobayashi H, Miyamae Y, Rojas E, Ryu JC, Sasaki YF (2000) Single cell gel/comet assay: guidelines for in vitro and in vivo genetic toxicology testing. Environ Mol Mutagen 35(3):206–221
29. Rahman A, Shahabuddin, Hadi SM, Parish JH, Ainley K (1989) Strand scission in DNA induced by quercetin and Cu(II): role of Cu(I) and oxygen free radicals. Carcinogenesis 10 (10):1833–1839
30. Czene S, Tibäck M, Harms-Ringdahl M (1997) pH-dependent DNA cleavage in permeabilized human fibroblasts. Biochem J 323(Pt 2):337–341
31. Gaggelli E, Kozlowski H, Valensin D, Valensin G (2006) Copper homeostasis and neurodegenerative disorders (Alzheimer's, prion, and Parkinson's diseases and amyotrophic lateral sclerosis). Chem Rev 106(6):1995–2044. doi:10.1021/cr040410w
32. Desai V, Kaler SG (2008) Role of copper in human neurological disorders. Am J Clin Nutr 88 (3):855S–858S
33. Madsen E, Gitlin JD (2007) Copper and iron disorders of the brain. Ann Rev Neurosci 30:317–337. doi:10.1146/annurev.neuro.30.051606.094232
34. Rinaldi AC (2000) Meeting report – Copper research at the top. Biometals 13(1):9–13. doi:10.1023/a:1009228824220
35. Pall HS, Williams AC, Blake DR, Lunec J, Gutteridge JM, Hall M, Taylor A (1987) Raised cerebrospinal-fluid copper concentration in Parkinson's disease. Lancet 2(8553):238–241
36. Lin C-N, Chen H-L, Yen M-H (2008) Flavonoids with DNA strand-scission activity from Rhus javanica var. roxburghiana. Fitoterapia 79(1):32–36. doi:10.1016/j.fitote.2007.07.008
37. Habtemariam S, Dagne E (2009) Prooxidant action of knipholone anthrone: copper dependent reactive oxygen species generation and DNA damage. Food Chem Toxicol 47(7):1490–1494. doi:10.1016/j.fct.2009.03.032
38. Pryor WA (1988) Why is the hydroxyl radical the only radical that commonly adds to DNA? Hypothesis: it has a rare combination of high electrophilicity, high thermochemical reactivity, and a mode of production that can occur near DNA. Free Radic Biol Med 4(4):219–223
39. Garcia-Garcia A, Zavala-Flores L, Rodriguez-Rocha H, Franco R (2012) Thiol-redox signaling, dopaminergic cell death, and Parkinson's disease. Antioxid Redox Signal 17(12):1764–1784. doi:10.1089/ars.2011.4501

40. Letelier ME, Sánchez-Jofré S, Peredo-Silva L, Cortés-Troncoso J, Aracena-Parks P (2010) Mechanisms underlying iron and copper ions toxicity in biological systems: pro-oxidant activity and protein-binding effects. Chem Biol Interact 188(1):220–227. doi:10.1016/j.cbi.2010.06.013
41. García CR, Angelé-Martínez C, Wilkes JA, Wang HC, Battin EE, Brumaghim JL (2012) Prevention of iron- and copper-mediated DNA damage by catecholamine and amino acid neurotransmitters, L-DOPA, and curcumin: metal binding as a general antioxidant mechanism. Dalton Trans 41(21):6458–6467. doi:10.1039/c2dt30060e

Chapter 8
Superconductivity in Human Body; Myth or Necessity

Athanasios Alexiou and John Rekkas

Abstract During the last years there is an increasing trend on the study of mitochondrial populations mainly in neural cells, due to their association with neurological disorders like Alzheimer's disease, Parkinson's disease, Autism, and CMT2A. Several studies concerning modeling of mitochondrial protein pathways, simulation of mitochondrial dynamics, biomarkers associated with Reactive Oxygen Species and many other related topics are already published. In this study we establish the idea of natural superconductivity in mitochondrial level as a necessary theoretical framework for the normal production of ATP and the avoidance of adverse reactions in Central Neural System.

Keywords Superconductivity • Mitochondrion • Inner mitochondrial membrane • Central neural system • Heavy metals • Reactive oxygen species • Neurological disorders • Bardeen–Cooper–Schrieffer theory • Cytochrome C • Ginzburg–Landau theory • Synchronistic phenomena • Entanglement correlations • Tunneling effect • Pauli exclusion principle

8.1 Introduction

Human body consists of several and different types of molecules which are electrically charged entities. While complex biochemical and electrical reactions release energy and transmit signals across the Central Neural System (CNS), the majority of adenosine triphosphate (ATP) synthesis seems to occur across energy-transducing membranes, like the Inner Mitochondrial Membrane (IMM). Besides the basic components such as Outer Mitochondrial Membrane (OMM), Mitochondrial Intermembrane Space (MIMS), and Matrix, IMM contains all the protein

A. Alexiou (✉)
Department of Informatics, Ionian University, Plateia Tsirigoti 7, Corfu 49100, Greece
e-mail: alexiou@ionio.gr

J. Rekkas
School of Science and Technology, Hellenic Open University, 18 Plateia Aristotelous, Patra 26335, Greece
e-mail: rekkas2004@yahoo.gr

P. Vlamos, A. Alexiou (eds.), *GeNeDis 2014*, Advances in Experimental Medicine and Biology 822, DOI 10.1007/978-3-319-08927-0_8

complexes and redox cofactors involved in electron transfer and ATP synthesis [1]. A Mitochondrial population or a Mitochondrial Culture [2] can be characterized as a dynamic set of interacting organelles with chaotic behavior due to their functions the so-called mitochondrial dynamics. While the number of mitochondria in a cell has to be regulated in order to satisfy cell's energy requirements, mitochondrial dynamics seems to play a functional role in maintenance of proper IMM electrical potential [3]. Latest researches have already associated mitochondrial alterations with a variety of symptoms including the nervous system, eyes, skeletal muscles, heart, liver, kidneys, and pancreas as well as neurodegeneration and dementia and the development of the most common diseases like Alzheimer's (AD), Parkinson's (PD), Huntington's (HD), Autism, and Charcot–Marie–Tooth disease type 2A (CMT2A). Additionally, mitochondria are also sites of formation of Reactive Oxygen Species (ROS), including Superoxide Anion [4] and the highly Reactive Hydroxyl Radical or its intermediates [5] as well as Reactive Nitrogen Species (RNS) such as Nitric Oxide [6]. Mitochondria generate endogenous ROS as by-products of oxidative phosphorylation [7]. Mitochondrial perturbations are known to participate in the mechanisms of human neuropathology, particularly disorders involving acute interruptions in O_2 and substrate delivery to the brain and bioenergetics failure as seen in tissue ischemia and toxic exposures [8, 9].

In this paper we describe a general biophysical framework for the theoretical concept of natural superconductivity occurrence in human body and especially in the IMM, adjusting and evaluating the necessity of merging Biology experiments with Quantum Biology assumptions and principles. While the Citric Acid Cycle or Krebs cycle as well as the function and structure of cytochrome c oxidase (COX) are already well described and referred in literature, we propose a new general model for the formulation of the simultaneous coexistence of Oxygen (O2), Iron (Fe), and Copper (Cu) and the electrons' transportation between mitochondrial Complex III and Complex IV. Our model is based on the application of the Bardeen, Cooper, and Schrieffer theory (BCS) [10] and the Ginzburg–Landau theory [11] as well as the occurrence of Synchronistic Phenomena due to Entanglement Correlations [12], the Tunneling Effect, and Pauli Exclusion Principle [13].

8.2 Natural Superconductivity into the Inner Mitochondrial Membrane

It is known that the most reliable way to measure the energy gap of a superconductor is the current-voltage curve in a metal/insulator/superconductor contact. For the mitochondrial case and between the complexes III and IV the iron (Fe) is the metal, the oxygen (O_2) is the insulator, and the copper (Cu) is the superconductor in the critical temperature (T_c) of the IMM (Fig. 8.1). Despite the insulating layer in the superconducting surface, electrons can be transferred from one side to another via the quantum tunneling effect. Superconductivity occurs due to the high levels of

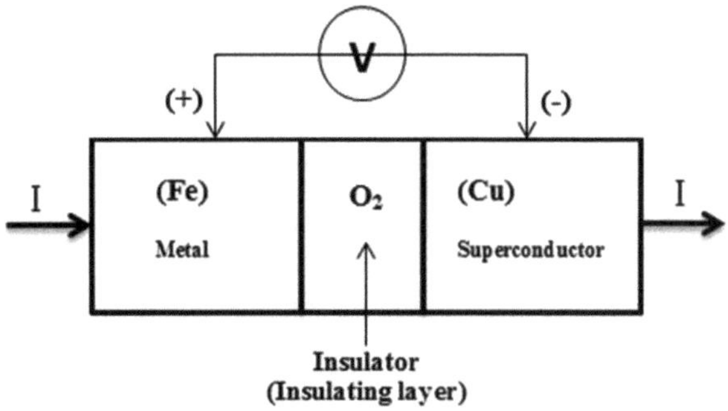

Fig. 8.1 Natural superconductivity between mitochondrial complex III and IV, at the IMM critical temperature (T_c)

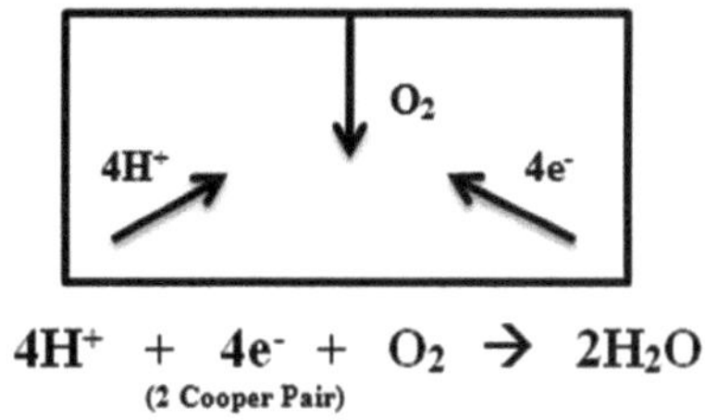

Fig. 8.2 H_2O production

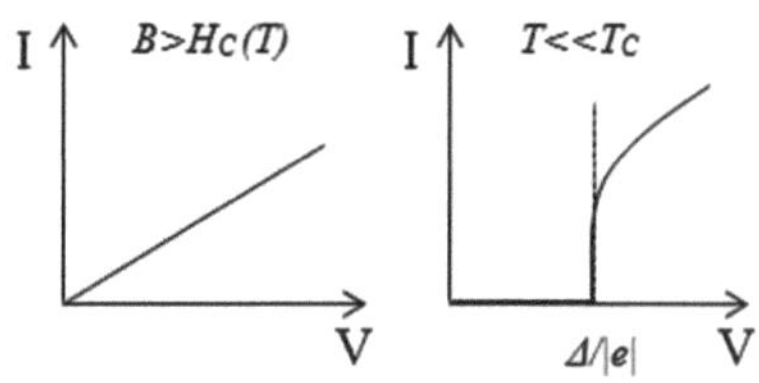

Fig. 8.3 Natural superconductivity in mitochondrial complex III

CuO_2 across the central axis of symmetry of the fundamental cell and in the orthorhombic crystal system.

Iron (Fe) binds oxygen and interacts with copper (Cu), creating a mixed oxide with a natural superconductivity at the critical mitochondrial temperature (T_c). Iron, Oxygen, and Copper create a superconducting surface where Iron and Copper will change their oxidation number. Iron and Copper oxidation numbers' increase or reduction will take place in order to switch from an initial superconductivity state to superconductivity destruction with simultaneous release of Oxygen.

There is a triple merge point in the superconducting surface, where (2) Oxygen atoms, four (4) electrons (e^-) forming two (2) Cooper pairs, and four (4) protons (H^+) from the mitochondrial matrix result in the formation of two (2) water molecules (H_2O) (Figs. 8.2 and 8.3).

In the 1st case the superconductor has reached the normal state (Superconductivity OFF) either by increasing the temperature above the critical (T_c) or by applying a specific magnetic field. In the 2nd case the superconductor remains in

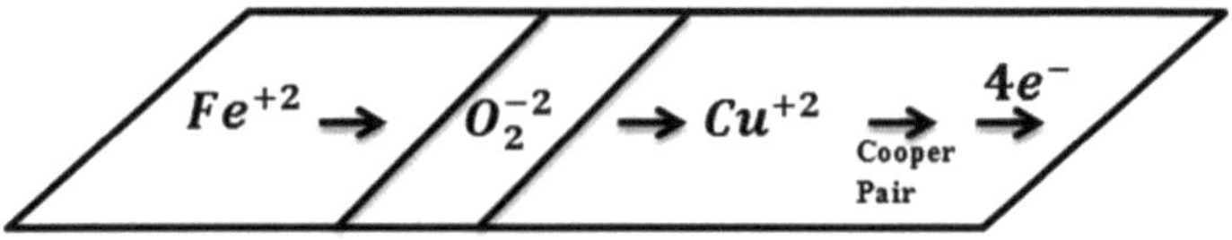

Fig. 8.4 Superconductivity switch (ON)

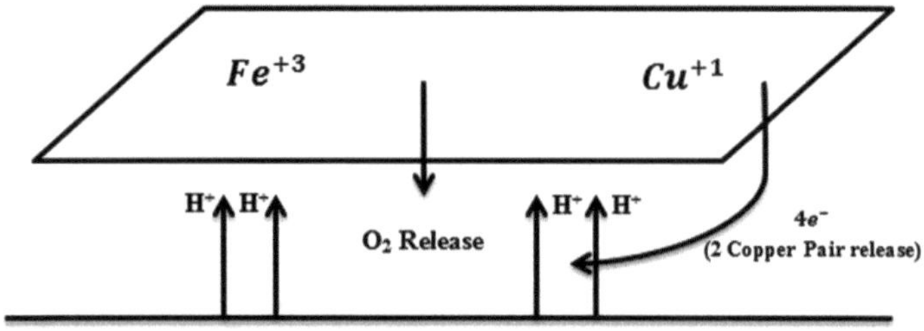

Fig. 8.5 Superconductivity switch (OFF)

the superconducting state (Superconductivity ON) and an electric potential flows through the merge point only when the voltage exceeds the value $\Delta/|e|$. A voltage below this limit is not conductive to electrons transfer due to the Pauli Exclusion Principle, in contrast with a simultaneously voltage increasing above this limit. Δ is the gap function according to the Ginzburg-Landau Theory.

In the initial state of the superconducting surface, two Cooper electron pairs are created due to the tunneling effect across the insulator. Then protons (H^+) are pumped from the mitochondrial matrix, approaching the superconductive surface resulting alteration in the electric field of the IMM surface and changes in the oxidation number of iron (Fe), from Fe^{+2} to Fe^{+3} and copper (Cu), from Cu^{+2} to Cu^{+1}.

Protons cannot pass over the boundary of the superconducting surface due to mutual Coulomb repulsion, between the (4) four protons (H^+) with the positive ions of iron (Fe^{+3}) and copper (Cu^{+1}). While the oxidation number of iron and copper is changed, the superconductivity surface is destroyed, releasing two Cooper pairs ($4e^-$) and oxygen (O_2).

The above process is repeated continuously with new (Fe), (O_2), and (Cu) atoms reaction. In each reaction the mechanism of the superconductive switch ON/OFF is activating (Figs. 8.4 and 8.5) in order to release O_2 and the 2 Cooper pairs ($4e^-$).

With this mechanism, we get O_2 release and not oxygen in the form of O_2^{-1} that may lead to the production of ROS. By creating two Cooper electron pairs we avoid the movement of single electrons which can bind to protons (H^+) derived from the mitochondrial trix, as well as the phenomenon of electrical thrombosis [3] and the establishment of further electrical complexes of (e^-) and (H^+). The existence of the above mechanism into the superconducting surface of Fe–O_2–Cu achieves the occurrence of normal respiratory chain in the IMM.

In order to establish our theory of natural superconductivity on human body and in subcellular scale we will address the following axioms based on certain preconditions of Biology and Quantum theory:

Axiom 1 (existence): Nowadays there are evidences concerning observed superconductivity far above the zero degrees Celsius on synthetic compounds [14]. The conditions of Fe–O_2–Cu coexistence in the IMM can lead to the development of local natural superconducting properties enhancing the function of the tunneling effect. A further characteristic and proof that copper in the IMM can act as a superconductive component is the fact that the required energy of copper's electrons is continuously renewed due to the observed piezoelectric effect during the mitochondrial fusion (Complete Fusion as a mitochondrial merge and Transient Fusion as a mitochondrial collision).

Axiom 2 (where): Superconductivity can occur only at subcellular level and specifically in the area of energy production into the IMM. It is not possible to observe superconductivity in other human body parts due to the interaction of several different types of membranes, cell layers, proteins, molecules, etc.

Axiom 3 (measurement): Due to the several human body synchronistic phenomena based on entanglement correlations, we can measure the impact of natural superconductivity using the tunneling phenomenon.

Obviously the observation, the recognition, and the evaluation of quantum phenomena in a human body still remain unknown, mainly when they refer to brain functionality and Central Neural System synchronism and disorders. We believe that either we refer to chemical and energy reactions or signal-information transmissions, tunneling phenomenon, and other well-formulated quantum assumptions have to be applied and be considered in Biology, Biomedicine, Cognitive Science, and Drug Delivery Systems.

Conclusion

Quantum theory and Modern Mathematics have already changed our way of comprehension of the universe. Since the beginning of the last century, many important and inspired theories had been established concerning the quantum phenomena, but until now very few have been applied to Biology. In this paper we examined the necessity of natural superconductivity occurrence in human body, as a result of combined known quantum principles and we focused our model on the mitochondrial inner membrane due to the co-existence of certain components.

References

1. Mannella CA (2006) The relevance of mitochondrial membrane topology to mitochondrial function. Biochim Biophys 1762:140–147

2. Alexiou A, Vlamos P (2012) A cultural algorithm for the representation of mitochondrial population, advances in artificial intelligence, vol. 2012, Hindawi Publishing, ID 457351, doi: 10.1155/2012/457351
3. Alexiou A, Rekkas J, Vlamos P (2011) Modeling the mitochondrial dysfunction in neurogenerative diseases due to high H+ concentration. Bioinformation 6(5):173–175, PubMed: 21738307
4. Liochev SI, Fridovich I (1995) Superoxide from glucose oxidase or from nitroblue tetrazolium. Arch Biochem Biophys 318:408–410
5. Halliwell B (2001) Role of free radicals in the neurodegenerative diseases. Drugs Aging 18:685–716
6. Zorov DB, Filburn CR, Klotz LO, Zweier JL, Sollott SJ (2000) Reactive oxygen species (ROS)-induced ROS release: a new phenomenon accompanying induction of the mitochondrial permeability transition in cardiac myocytes. J Exp Med 192(7):1001–1014
7. Wallace DC (2005) A mitochondrial paradigm of metabolic and degenerative diseases, aging, and cancer: a dawn of evolutionary medicine. Annu Rev Genet 39:359–407
8. Isaev NK, Stel'mashuk EV, Zorov DB (2007) Cellular mechanisms of brain hypoglycemia. Biochemistry (Mosc) 72(5):471–478, Review
9. Kirpatovskii VI, Kazachenko AV, Plotnikov EY, Kon'kova TA, Drozhzheva VV, Zorov DB (2007) Effects of ischemic and hypoxic preconditioning on the state of mitochondria and function of ischemic kidneys. Bull Exp Biol Med 143(1):105–109
10. Bardeen J, Cooper* LN, Schrieffer JR (1957) Theory of superconductivity. Phys Rev 108 (5):1175–1204
11. Ginzburg VL, Landau LD (1950) Zh Eksp Teor Fiz 20:1064
12. Atmanspacher H, Roemer H, Walach H (2001) Weak quantum theory: complementarity and entanglement in physics and beyond, arXiv:quant-ph/0104109. J Ref Found Phys 32 (2002):379–406
13. Pauli W (1925) On the connexion between the completion of electron groups in an atom with the complex structure of spectra. Z Phys 31:765
14. Website: http://www.superconductors.org/65C.htm

Chapter 9
Molecular Genetics of Huntington's Disease

Catherine Bobori

Abstract Huntington's disease is a neurodegenerative disorder of the brain that is caused by the mutation of the gene which produces a protein called huntingtin (htt). The mutation is based on the continuous repetition of the trinucleotide CAG which in turn makes the protein toxic for the brain cells. As a result neurons which contain the mutant protein begun to atrophy. The loss of those brain cells can cause many problems to the patients, even death. The aim of this paper is to report the problems caused to the brain by the mutant protein, specifically in the area of basal ganglia, the area that is the most affected by the disease, as well as the symptoms and metabolic changes to which the patient is subjected. Finally, the summary of the methods for a more timely and accurate diagnosis of the disease is based on these changes in order to simplify and facilitate the lives of inflicted people by means of the administration of the appropriate treatment.

9.1 Introduction

Huntington's disease (also known as Huntington's chorea) is an inherited and neurodegenerative disorder of the brain, which is focused on the region of the basal ganglia and is causing mental and emotional disorders.

The disease is caused due to an autosomal mutation of both copies (autosomal dominant inheritance) on chromosome 4. This gene produces the protein huntingtin (htt). The produced defective protein leads to changes in the brain that cause abnormal, involuntary movements (facial and motor), reduced thinking ability, irritability, anxiety, depression, obsessive behavior, and other mood changes. The disease destroys nerve cells (neurons) in certain parts of the brain. As a consequence there is a reduction of neurotransmitters that carry signals to the brain. The disease can cause death after about 10–25 years. The molecular basis of the disease is the expansion of triplet nucleotide repeats of CAG in exon 1 of the IT15 gene on

C. Bobori (✉)
Department of Informatics, BiHELab, Ionian University,
Plateia Tsirigoti 7, Corfu 49100, Greece
e-mail: p12bobo@ionio.gr

P. Vlamos, A. Alexiou (eds.), *GeNeDis 2014*, Advances in Experimental Medicine and Biology 822, DOI 10.1007/978-3-319-08927-0_9

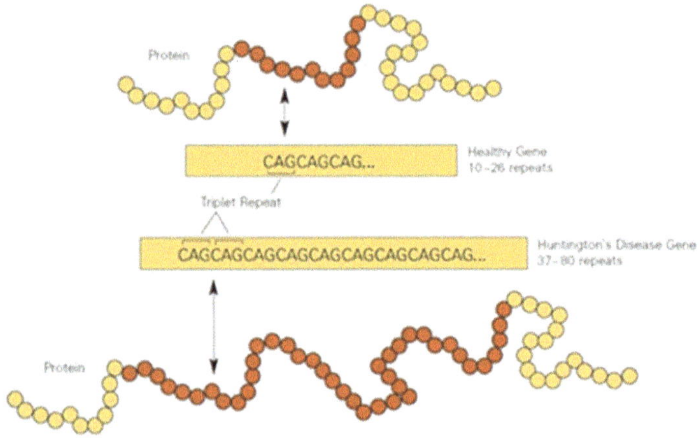

Fig. 9.1 This image shows the repetition of the trinucleotide CAG in the mutant gene. In HD a person carries 39–70 triplets in at least one chromosome

chromosome 4p16.52. Physiologically alleles carry about 8–30 triplets, while the inflicted ones carry approximately 39–75 triplets in at least one chromosome (Fig. 9.1). The mutation results in the longitudinal stretching of glutamine near the NH_2 terminus of the protein. Symptoms occur primarily at 30–50 years of age, but there are cases that appear earlier during childhood or adolescence. In addition, the symptoms differ from person to person depending on the extent of the mutation. The prevalence of the mutation is 4–10 cases per 100,000 population in Western Europe, with many more to be at greater risk of inheriting the mutated gene (Table 9.1).

The purpose of this paper is to present the changes taking place in the body of the patient on account of the Huntington's disease, primarily in the nerve cells. Finally, potential approaches to a more timely diagnosis of the disease will be presented based on symptoms and metabolic changes that the patient is experiencing.

9.2 Effects of Huntington's Disease on the Basal Ganglia

The basal ganglia are located at the base of the forebrain and are composed of the striatum (caudate and putamen) and the globus pallidus. The pathology of Huntington's disease is concentrated on a selected neuronal population (caudate nucleus and neurons) while places such as the shell and the cerebral cortex are less affected (Fig. 9.2). The cerebellum remains in good condition.

Almost all motor and sensory nerve fibers through which the shell is linked to the spinal cord are passed through the caudate nucleus and the shell.

In this disease we observe an excessive activity of the direct pathway (the activity of cerebral cortex that stimulates cells in the striatum, which cells participate in the direct pathway, leads to the inhibition of areas GPi and SNr, which in

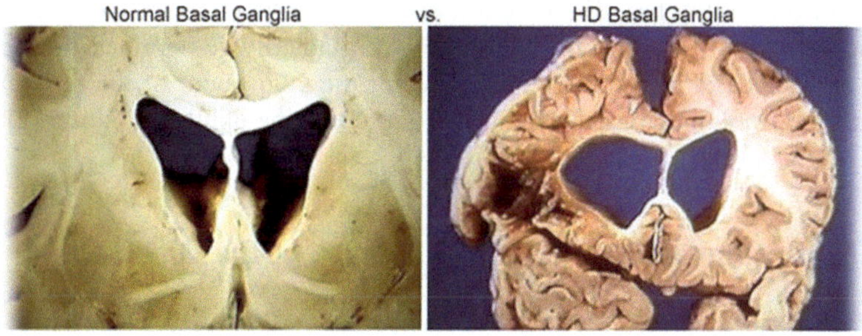

Fig. 9.2 This picture represents the impact of HD on the basal ganglia of the human brain. It is important to note that the brain of a person with HD has bigger openings due to the atrophy that occurs in the neurons

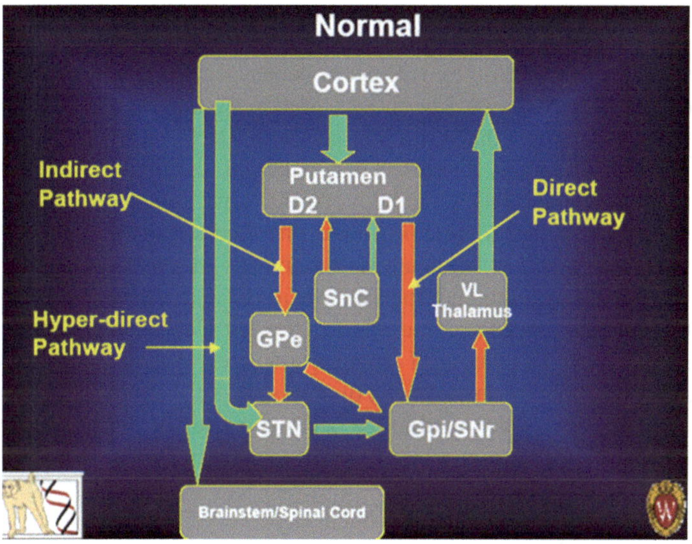

Fig. 9.3 This diagram presents the combination of the function of both the direct (Putamen > GPi/SNr > Thalamus > Cortex) and indirect (Putamen > GPe > GPi/SNr > Thalamus > Cortex) pathway of brain's basal ganglia. In Huntington's disease we observe an excessive activity of the direct pathway while there is an insufficient activity in the indirect pathway

turn remove the tonic inhibitory activity toward the thalamus). Moreover, there is an insufficient activity in the indirect pathway (the activity of cerebral cortex that stimulates cells in the striatal indirect pathway is considered to inhibit the thalamus (blocking the lifting of the suspension)) (Fig. 9.3). The striatum shows the greatest atrophy during the disease which results in the patient's incapability to perform simple movements. Due to the destruction of the motor cortex the patient lacks the ability to perform fine dexterous movements but without losing the ability to perform massive prominent movements. The destruction of the striatum leads to the loss of this ability too. The function of basal ganglia is associated with actions that the person suffering from the disease will gradually be deprived of, too. The motor loop of basal ganglia is responsible for the adjustment of the upper motor neurons, the smooth commencement of motion, and the ophthalmic loop [2].

9.3 Symptoms and Molecular Changes of the Disease

Huntington's chorea primarily affects the basal ganglia of the brain leading to atrophy and reduction of brain nerve cells. As a result, the patient suffers from mental disorders such as depression, irritability, anxiety, compulsive behavior, and violence. In addition, at the early stages patients suffer from chorea and dystonia as well as inability to recall memory, reduced capability of thinking, and reduced executive function. Symptoms occur primarily at 30 to 50 years of age; however there are cases that begin to appear during childhood-adolescence. Severity of symptoms as well as when they will occur depend on the extent of the mutation and the degree of heritability. The mutant protein (m-htt) becomes toxic to brain cells. Surveys have shown that m-htt interacts with Rhes which is found only in the striatum and this is probably the reason that makes htt toxic only to nerve cells even though it is found in all cells in the body. At the same time, the m-htt inhibits the ability of normal protein to perform molecular interactions that are crucial for the survival and function of neurons that mostly degenerate in the disease. Also the excitotoxicity holds an important role in HD and depends on the excitation of NMDA glutamate receptors, the internal flow of Ca^{2+}, and the production of NO. The metabolism of glucose of striatum and cerebral cortex is reduced in HD and appears to precede the loss of most of the tissue. Furthermore, increased levels of lactic acid have been identified in the occipital cortex, and it has been recently discovered that the expansion of CAG repetition is associated with the levels of lactic acid in the striatum.

Research results indicate that mitochondria are very likely to be associated with neurodegenerative diseases that cause aging, as mitochondria are the main regulators of the cell cycle, apoptosis, and homeostasis of Ca^{2+}. In all neurodegenerative diseases, and in HD, there is strong evidence that mitochondrial dysfunction occurs at the early stages and acts causally in the pathogenesis of the disease. Research has shown that in patients with HD mitochondria depolarize lower loads of calcium and their membranes have a lower potential. The same defect seems to exist in mice which have a full-length mutation [3, 6, 8–10, 12, 14]. The decreased striatal metabolism that occurs in presymptomatic HD patients shows that striatal neurons are particularly vulnerable to disruptions of mitochondrial functions Some researchers have shown that increased PGC-1α can reduce the influence of m-htt in mitochondria. Furthermore, it is believed that when the concentration of PGC-1α is increased, it helps protect neurons against toxicity of m-htt while on the contrary when its concentration is decreased the nerve cells are vulnerable to mutant protein. PGC-1α is a transcriptional coactivator which regulates several metabolic

Table 9.1 The table shows what is the change of the protein and whether the person is suffering from the disease depending on the extent of the trinucleotide repetition

Classification of the trinucleotide repeat, and resulting disease status, depends on the number of CAG repeats [13]			
Repeat count	Classification	Disease status	Risk to offspring
<26	Normal	Will not be affected	None
27–35	Intermediate	Will not be affected	Elevated but <50 %
36–39	Reduced change of protein	May or may not be affected	50 %
40+	Full change of protein	Will be affected	50 %

processes, including mitochondrial biogenesis and oxidative phosphorylation. It seems that interaction of TAF4 and CREB results in the significant protection of striatal cells and PGC-1α from the effects of the mutant protein. It is observed that the mutant huntingtin inhibits transcription of the PGC-1 by interfering with the action of CREB/TAF4 in the promoter of PGC-1α (Table 9.1) [7].

9.4 Modern Approaches of Huntington's Disease

The knowledge we have gained at the molecular level about the mechanisms that generate genetic diseases has given us the opportunity to develop methods for their detection. Early diagnosis of a genetic disease provides us with the ability to schedule the best treatment in order to reduce the complications of the disease. Huntington's disease like many other diseases that are caused by mutations can be detected with conventional diagnostic techniques of traditional medicine.

First of all, there is prenatal diagnosis for the disease. If a couple who suffers from the disease wishes to have children, they should turn to either chorionic villus sampling or amniocentesis test. Embryonic cells can be used to analyze the DNA and to detect or not the repetition of the three nucleotides. For those who may have inherited the disease but have not yet experienced symptoms (which is the case of one-parent sufferer) it is appropriate to take the presymptomatic test for DNA analysis and detection of mutant protein. Finally a confirmatory test for people who are beginning to show symptoms of the disease includes psychological testing and detection of the gene. The psychological tests examine whether people obtain violent behavior, irritability, anxiety, depression, and so on. The extent of the repetition could be detected with various biochemical processes and analysis of DNA's sequence. Additionally, by using a magnetic scanner and a CT scanner the doctor can take pictures of the brain to diagnose if there are spots in the brain that have begun to atrophy (degeneration may begin much earlier than the onset of symptoms).

Besides these traditional methods I would like to add some other points that may lead us to an early diagnosis of the disease based on the changes that occur in the body of the patient at the molecular level. To begin with, since there are changes that seem to occur in the mitochondria, through various biochemical processes we could observe the capacity of Ca^{2+} in the cytoplasm of the organelle. The patient could be examined and observed for any changes that may occur in the membrane of the organelle as well. Another possible tactic is to monitor the levels of PGC-1α and the CREB/TAF4 interacting with it.

Conclusion

In summary, the disease causes severe biochemical changes to specific substances and leads as well to the atrophy of the nerve cells of the brain. The metabolic changes that occur to the patients can show to us the molecular mechanisms of the disease.

Based on the surveys conducted so far there is no proven and successful treatment for Huntington's disease. Doctors usually suggest some treatment to reduce and control convulsive movements, psychological problems, and memory loss. This treatment is trying to reduce the mental disorders in which the patient is subjected to. However, the pharmaceutical treatment will be more effective if the administration starts at the early stages of the disease. That is why an early diagnosis is really important. By studying the way our body reacts to the disease we can understand its effects on the patients' lives and try to deal with the disease and its difficulties [1].

Hopefully, with the techniques mentioned above, early and accurate diagnosis is possible in order to offer patients a better and easier life [11].

References

1. Zuccato C, Valenza M, Cattaneo E (2010) Molecular mechanisms and potential therapeutical target in Huntington's disease. Department of Pharmacological Sciences and Centre for Stem Cell Research, Universita' degli Studi di Milano, Milan, Italy
2. Laforet GA, SappE, Chase K, McIntyre C, Boyce FM, Campbell M, Cadigan BA, Warzecki L, Tagle DA, Hemachandra Reddy P, Cepeda C, Calvert CR, Jokel ES, Klapstein GJ, Ariano MA, Levine MS, DiFiglia M, Aronin N, Changes in Cortical and Striatal Neurons Predict Behavioral and Electrophysiological Abnormalities in a Transgenic Murine Model of Huntington's Disease (2001) Departments of Psychiatry and Medicine, University of Massachusetts Medical School, Worcester, MA, Department of Neurology, Massachusetts General Hospital, Boston, MA, Genetics and Molecular Biology Branch, National Genome Research Institute, National Institutes of Health, Bethesda, MD, Mental Retardation Research Center, University of California at Los Angeles, Los Angeles, CA, and Department of Neuroscience, Chicago Medical School, North Chicago, IL, J Neurosci 21(23):9112–9123
3. Maria Damiano, Laurie Galvan,Nicole Diglon, Emmanuel Brouillet, Mitochondria in Huntington's disease, Elsevier 11 August 2009
4. Mitochondria in Huntington's disease CEA, DSV, I2BM Molecular Imaging Research Center (MIRCen), CEA, CNRS URA 2210, F-92265 Fontenay-aux-Roses, France, 31 July 2009
5. Yamamoto A†, Lucas JJ†‡, Hen R (2000) Reversal of neuropathology and motor dysfunction in a conditional model of Huntington's disease, Center for Neurobiology and Behavior Columbia University New York, New York, Cell, vol. 101, pp 57–66

6. Cunha-Oliveira T, Ferreira IL, Cristina Rego A. Consequences of mitochondrial dysfunction in Huntington's disease and protection via phosphorylation pathways, CNC-Center for Neuroscience and Cell Biology, University of Coimbra, Faculty of Medicine, University of Coimbra, Portugal
7. Cui L, Jeong H, Borovecki F, Parkhurst CN, Tanese N, Krainc D (2006) Transcriptional repression of PGC-1a by mutant Huntingtin leads to mitochondrial dysfunction and neurodegeneration, Department of Neurology, Massachusetts General Hospital, Harvard Medical School, MassGeneral Institute for Neurodegeneration, Charlestown, MA, USA, Department of Microbiology, New York University School of Medicine, New York, NY, Cell, vol 127, pp 59–69, Elsevier Inc.
8. Mandemakers W, Morais VA, De Strooper B. A cell biological perspective on mitochondrial dysfunction in Parkinson disease and other neurodegenerative diseases, Center for Human Genetics, K. U. Leuven and Department of Molecular and Developmental Genetics, VIB, Herestraat 49, 3000 Leuven, Belgium, Accepted 13 March 2007 J Cell Sci 120:1707–1716 Published by The Company of Biologists 2007 doi:10.1242/jcs.03443
9. Rego C, Oliveira CR (2003) Mitochondrial dysfunction and reactive oxygen species in excitotoxicity and apoptosis: implications for the pathogenesis of neurodegenerative diseases. Neurochem Res 28(10):1563–1574
10. Baron M, Kudin AP, Kunz WS (2007) Mitochondrial dysfunction in neurodegenerative disorders, Department of Epileptology and Life & Brain Center, University Bonn, Sigmund-Freud-Strasse 25, D-53105 Bonn, Germany, Biochemical Society Transactions, vol 35, part 5
11. Alexiou A, Psixa M, Vlamos P (2011) Ethnical issues of artificial biomedical applications, Department of Informatics, Ionian University, Corfu, Greece, Artificial Intelligence Applications and Innovations IFIP Advances in Information and Communication Technology, vol 364, pp 297–302
12. Alexiou A, Rekkas J, Vlamos P (2011) Modeling the mitochondrial dysfunction in neurogenerative diseases due to high H+ concentration. Bioinformation 6(5):173–175, PubMed: 21738307
13. Alexiou A, Vlamos P (2012) Evidence for early identification of Alzheimer's disease. ArXiv:1209.4223v2 [q-bio.NC]
14. Alexiou A, Vlamos P (2012) A cultural algorithm for the representation of mitochondrial population, advances in artificial intelligence, vol. 2012, Hindawi Publishing, ID 457351, doi: 10.1155/2012/457351

Chapter 10
Altered Galectin Glycosylation: Potential Factor for the Diagnostics and Therapeutics of Various Cardiovascular and Neurological Disorders

Ghulam Md Ashraf, Asma Perveen, Shams Tabrez, Nasimudeen R. Jabir, Ghazi A. Damanhouri, Syed Kashif Zaidi, and Naheed Banu

Abstract Galectins are β-galactoside binding mammalian proteins characterized by the presence of a conserved carbohydrate recognition domain, expressed in almost all taxa of living organisms and involved in broad range of significant biological and physiological functions. Previously, we reported the purification and extensive characterization of galectin-1 from goat (*Capra hircus*) heart. Interestingly, the purified protein was found to have significant level of glycosylation. This intrigued us to evaluate the involvement of glycosylation in relation to protein's structural and functional integrity in its purified form. In the present study, an extensive comparative physicochemical characterization has been performed between the glycosylated and deglycosylated form of the purified protein. Deglycosylation resulted in an enhanced fluorescence quenching and marked reduction in pH and thermal stability of the purified galectin. Exposure to various biologically active chemicals showed significant differences in the properties and stability profile, causing significant deviations from its regular secondary structure in the deglycosylated form. These results clearly indicated enhanced structural and

G.M. Ashraf (✉) • S. Tabrez • N.R. Jabir • G.A. Damanhouri
King Fahd Medical Research Center, King Abdulaziz University,
P.O. Box 80216, Jeddah 21589, Saudi Arabia
e-mail: ashraf.gm@gmail.com; gashraf@kau.edu.sa

A. Perveen
Department of Biochemistry, Faculty of Life Sciences, Aligarh Muslim University,
Aligarh 202002, Uttar Pradesh, India

S.K. Zaidi
Center of Excellence in Genomic Medicine Research, King Abdulaziz University,
P.O. Box 80216, Jeddah 21589, Saudi Arabia

N. Banu (✉)
Department of Biochemistry, Faculty of Life Sciences, Aligarh Muslim University,
Aligarh 202002, Uttar Pradesh, India

College of Medical Rehabilitation, Qassim University, Buraydah, Saudi Arabia
e-mail: naheedbanu7@yahoo.com; nbanuamu@gmail.com

P. Vlamos, A. Alexiou (eds.), *GeNeDis 2014*, Advances in Experimental
Medicine and Biology 822, DOI 10.1007/978-3-319-08927-0_10

functional stabilization in the glycosylated galectin. The data revealed herein adds a vital facet demonstrating the significance of galectin expression and glycosylation in causation, progression, and possible therapeutics of associated clinical disorders. Our approach also allowed us to define some key interactions between the purified galectin and carbohydrate ligands that could well serve as an important landmark for designing new drug protocols for various cardiovascular and neurological disorders.

Keywords Goat heart galectin-1 (GHG-1) • Glycosylation • Deglycosylation • Spectroscopy • Cardiovascular disorders • Neurological disorders

10.1 Introduction

Galectins are a family of lectins characterized by the presence of at least one carbohydrate recognition domain (CRD) with an affinity for β-galactosides and sharing certain conserved sequence elements [1, 2]. Members of the galectin family are found in mammals, fish, birds, amphibians, sponges, nematodes, and some fungi [3], and are well known to perform intracellular as well as extracellular functions [4]. Of the 15 galectins reported till date, galectin-1 (Gal-1) and galectin-3 (Gal-3) are the most reported members of galectin family, typified by its marked affinity for β-galactosides and a conserved 134 amino acid CRD, and involved in wide-ranging gamut of vital biological functions [3, 5]. Galectins have been reported to be omnipresent in all animal tissues including mammalian heart and brain [3, 6]. In heart, galectins are mainly localized in connective-tissue elements, vascular structures, myocardial cell constituents, and endocardial tissue [2, 7]. Varied galectin expression has been shown to be indigenously involved in causation and progression of different cardiovascular disorders like hypoxia-induced pulmonary hypertension, Chagas' disease, atherosclerosis, and restenosis [2, 8].

In brain, galectins have been reported to be localized in ependymal cells, ionized calcium-binding adapter 1-positive macrophages/activated microglia, and some astrocytes [9]. Several galectins have been reported with altered expression during brain tumor progression, thus making them potential targets for the diagnosis and treatment of brain cancer [10]. Moreover, altered galectin expression has been reported to be in various neurodegenerative disorders. Gal-1 prevents neurodegeneration and promotes neuroprotection in the brain [11]. Gal-1 has also been reported to protect from inflammation-induced neurodegeneration by deactivating classically activated microglia [12]. Gal-3 has been reported to play an important role in the subsequent neurodegeneration in murine scrapie [13]. Elevated level of serum Gal-3 has been reported in AD patients, making it a potential biomarker for AD [14]. Elevations in Gal-3 by microglia with progression of chronic neurodegenerative diseases represent a protective, anti-inflammatory innate immune response to chronic motor neuron degeneration [15].

Glycosylation has been reported to be one of the most important post-translational modifications for newly synthesized proteins having great impact on biological as well as physicochemical functions [6, 16]. Glycosylation of unfolded proteins and its involvement in the quality control of glycoprotein congregation in the endoplasmic reticulum is quite evident [17]. Although galectins are frequently expressed on cell surfaces or in extracellular matrix, they often lack familiar secretion signal sequences and do not follow the typical Golgi/endoplasmic reticulum pathway [18]. Consequently, majority of the galectins have features similar to the cytoplasmic proteins, for instance having free sulfhydryl groups, an acetylated N-terminus, and lack of glycosylation. Galectins are thus habitually looked upon as non-glycosylated proteins. However, an interesting exception was observed when we performed carbohydrate content analysis of goat heart galectin-1 (GHG-1 [8]). We found this galectin to be significantly glycosylated. This exciting finding in its crude form was worth exploration, for the possible connotation of galectin glycosylation with various cardiovascular and neurodegenerative disorders. Our claim is further strengthened by the finding that Gal-1–glycan interactions are essential in tempering microglia activation, brain inflammation, and neurodegeneration with critical therapeutic implications for multiple sclerosis (MS) [12]. Moreover, it was also reported that in neuroblastoma cells, Gal-1 is involved in regulating cell surface retention of RPTPβ and its phosphatase activity through O-mannosyl glycans synthesized by GnT-Vb [19]. In another study, it was reported that elevated serum Gal-3 levels might suggest potential proapoptotic activation, inflammation, and impaired neurodegeneration in patients with AD, and have been suggested as a potential biomarker for AD [14]. Citing the involvement of galectins and the role of galectin glycosylation in various cardiovascular and neurological disorders, we investigated the effect of glycosylation on the properties of GHG-1 to explore associated implications on disease progression. The findings of this study is expected to provide new insights in galectin biology.

10.2 Materials and Methods

10.2.1 Reagents

Sephadex G_{50}, sephadex G_{100}, proteins, periodic acid, SDS (sodium dodecyl sulfate), EDTA (ethylene diamine tetra acetic acid), tris, saccharides, bromophenol blue, phenol-sulfuric acid, coomassie brilliant blue G-250 and R-250 were purchased from sigma chemical co. (USA). Rest of the chemicals used were of the analytical grade. All the experimental procedures and protocols were repeated at least three times, and the average data has been reported in this manuscript.

10.2.2 Ethics Statement

All animal experimentations were permitted by Ministry of Environment and Forests, Government of India under registration no 714/02/a/CPCSEA issued by Committee for the Purpose of Control and Supervision of Experiments on Animals (CPCSEA) dated 16th November, 2002. All experiments on animals were approved by Departmental Ethical Committee (acad/D-833/ ILK/07-09-2007), Department of Biochemistry, AMU, India.

10.2.3 Purification of Goat Heart Galectin-1 (GHG-1)

The GHG-1 was isolated and purified in essence according to the methods described by Ashraf et al. [2]. The protein concentration was estimated by Lowry et al. [20] and its activity was determined by hemagglutination activity using trypsinized rabbit erythrocytes by two fold serial dilutions on microtiter V-shaped plate (Laxbro, India) [21].

10.2.4 Estimation of Degree of Glycosylation in GHG-1

The degree of glycosylation in the purified protein was determined by Dubois method [22] and the results were confirmed by PAS staining [23].

10.2.5 Deglycosylation of Glycosylated GHG-1 (gGHG-1)

The method described by Rasheedi et al. [24] was followed to carry out deglycosylation of the gGHG-1. The protein sample (1 mg/ml) was dissolved in PBS "B" [75 mM sodium phosphate pH 7.2, containing 0.15 M NaCl, 5 mM β-ME, and 0.02 % (w/v) sodium azide]. The solution was then acted upon with 10 mM sodium periodate solution in 5:1 molar ratio and incubated at room temperature in dark for 15 min. Oxidation was stopped by treating the reaction mixture with 0.5 ml ethylene glycol per ml protein sample. This was followed by dialysis of the quenched sample against PBS "B" at room temperature overnight. Deglycosylation was confirmed by Dubois method [22], PAS staining [23], and comparative mobility of gGHG-1 and deglycosylated GHG-1 (dgGHG-1) on SDS-PAGE [25].

10.2.6 Post-deglycosylation Changes in Electrophoretic Pattern and Molecular Weight

The changes in electrophoretic pattern and molecular weight of dgGHG-1 were monitored by SDS-PAGE by the method of Weber and Osborn, where the relative migration and mobility of dgGHG-1 were compared to that of gGHG-1 [25]. The change in the molecular weight of dgGHG-1 compared to gGHG-1 was also investigated under native (without β-ME) and reduced (with β-ME) environment by sephadex G_{100} gel permeation chromatography.

10.2.7 Post-deglycosylation Changes in Stokes Radius

Stokes radii of gGHG-1 and dgGHG-1 were calculated by gel filtration data. The partition coefficient (K_{av}) of every molecular weight marker protein was calculated by the below mentioned formula:

$$K_{\mathrm{av}} = (V_{\mathrm{e}} - V_{\mathrm{o}})/(V_{\mathrm{t}} - V_{\mathrm{o}})$$

where V_e is "elution volume," V_t is "total bed volume," V_o is "void volume." A linear plot between square root of the $-\log[K_{av}]$ and stokes radii of marker proteins was used for calculating Stokes radii of gGHG-1 and dgGHG-1.

10.2.8 Post-deglycosylation Changes in Diffusion Coefficient

The diffusion coefficients (D) for gGHG-1 and dgGHG-1 corresponding to their Stokes radii were calculated with the help of Stokes-Einstein equation:

$$D = KT/6\pi\eta r$$

where K is "Boltzman constant (1.386×10^{-16} erg/deg)," T is "absolute temperature (303 k)," η is "coefficient of viscosity of the medium (0.0100 P for water) and dilute aqueous salt solutions at 20 °C," r is "Stokes radius."

10.2.9 Post-deglycosylation Modulation in Thermal and pH Stability

The gGHG-1 and dgGHG-1 proteins (100 μg/ml each in PBS "B") were incubated for 30 min at a temperature range of 30–80 °C. Similarly, for pH stability monitoring, the gGHG-1 and dgGHG-1 proteins (100 μg/ml each in PBS "B" containing

50 μl of normal saline and 5 mM β-ME) were incubated for 24 h at 4 °C with 50 ml of following buffers: 0.1 M sodium acetate buffer (pH 3.5–5.5), 0.1 M sodium phosphate buffer (pH 6.5–7.5), 0.1 M Tris–HCl buffer (pH 8.5–9.5), and 0.1 M glycine-NaOH buffer (pH 10.5–11.5). Every sample was then examined for hemagglutination activity [21].

10.2.10 Deglycosylation-Mediated Inflection in Reaction Pattern of Various Biologically Active Reagents

The effect of thiol-blocking agents was monitored by incubating gGHG-1 and dgGHG-1 (100 μg/ml each in PBS "B") with 70 mM each of iodoacetamide, *p*-hydroxymercuribenzoate (*p*HMB), and N-ethylmaleimide (NEM) in an increasing incubation period (0–60 min) at room temperature. The effect of various denaturants was determined by incubating gGHG-1 and dgGHG-1 (100 μg/ml each in PBS "B") with increasing concentrations (0–8 M) of urea, guanidine HCl (GdnHCl), and thiourea at 37 °C for 1 h. The effect of different detergents was monitored by incubating gGHG-1 and dgGHG-1 (100 μg/ml each in PBS "B") with increasing concentrations (1–5 M) of SDS, Tween-20, and Triton X-100 at 37 °C for 1 h. After assigned times, all the samples were observed for hemagglutination activity [21].

10.2.11 Spectroscopic Studies

UV, fluorescence, far UV-CD, near UV-CD, and FTIR spectra were carried out according to the methods described in Ashraf et al. [6].

10.3 Results and Discussion

In our earlier study, we purified a galectin-1 from goat (*Capra hircus*) heart tissue, and carried out its characterization, structural analyses, and bioinformatics studies [8]. As an extension to the physicochemical studies, just as a matter of chance, we performed Dubois analysis [22] to check galectin glycosylation, if any. We were pleasantly surprised to note the presence of approx. 4.25 % sugar moiety of the total weight of GHG-1. Initially, we considered it as a false positive result, which might possibly have been due to contamination induced while testing GHG-1 affinity with various saccharides. To rule out these possibilities, we carried out extensive dialysis of the protein sample against PBS "B," to remove any saccharide contamination. But, Dubois analysis again revealed approx. 4.25 % sugar moiety of the total weight of GHG-1. The same procedure was repeated many times to validate this significant finding, and was further authenticated by PAS staining [23] of the electrophoresis gels, and deglycosylation studies discussed elsewhere in this manuscript.

Available literatures suggest that the addition of glycan structures to the protein backbone can radically alter its structure and function [26]. Linked glycans can influence the protein structure in two ways: (a) addition of glycans to the partially folded nascent polypeptide can impact/facilitate the protein folding pathway, and (b) the glycans can stabilize the mature protein. The revelation of GHG-1 being remarkably glycosylated encouraged us to examine the potentially anticipated role of glycosylation on the structure and function of GHG-1 by a cooperative analysis of its structural stability and activity in glycosylated as well as deglycosylated forms. The GHG-1 was deglycosylated by periodate oxidation method [24]. Dubois analysis [22], PAS staining [23], and the difference in the mobility of gGHG-1 and dgGHG-1 on SDS-PAGE [25] confirmed the deglycosylation results.

On Sephadex G_{100} gel permeation column, dgGHG-1 eluted earlier than gGHG-1; the dgGHG-1 eluted as a monomer of 14.1 kDa (compared to 14.5 kDa of gGHG-1) under reducing conditions and as a dimer of 28.2 kDa (compared to 29 kDa of gGHG-1) under nonreducing conditions. On SDS-PAGE, dgGHG-1 eluted earlier than gGHG-1, which correspond to its enhanced mobility on the gel (Fig. 10.1). The decreased molecular weight of the dgGHG-1 compared to gGHG-1 seems to be direct consequence of deglycosylation. This finding suggests that galectin dimerization is not perturbed by deglycosylation, thus establishing the fact that the glycans detached during deglycosylation are not involved in the dimerization of GHG-1 subunits.

Stokes radius has been defined as the average radius of a protein in solution, and depends on its 3D molecular structure. Practically, purification of a protein also depends on its Stokes radius, rather than directly on its molecular weight. The data generated during gel filtration chromatography were used to compute the Stokes radii. Post-deglycosylation, Stokes radius of the monomeric form of the protein decreased from 17.5 Å for gGHG-1 to 16.4 Å for dgGHG-1; and that of the dimeric form decreased from 26.0 Å for gGHG-1 to 24.9 Å for dgGHG-1 (Table 10.1). The decreased Stokes radii of monomeric as well as dimeric dgGHG-1 proteins further confirmed deglycosylation, as the removal of glycan moieties by deglycosylation decreased the average radius (Stokes radius) of dgGHG-1. The reduced Stokes radius of dgGHG-1 allowed its faster passage through polyacrylamide and sephadex gel matrices, thus further justifying the enhanced mobility of dgGHG-1 on SDS-PAGE and gel filtration column.

The diffusion coefficients were calculated to estimate the number of water molecules bound to the purified protein [19]. Post-deglycosylation, diffusion coefficient of the monomeric form of the protein increased from 12.7×10^{-15} cm^2/s for gGHG-1 to 13.86×10^{-15} cm^2/s for dgGHG-1; and that of the dimeric form increased from 8.57×10^{-15} cm^2/s for gGHG-1 to 9.71×10^{-15} cm^2/s for dgGHG-1 (Table 10.1). Deglycosylation eliminates the sugar moieties attached to the protein surface, thus allowing increased access of the water molecules to the protein surface. As the number of bound water molecules decrease, the mobility of the protein–water complex increases, along with the diffusion coefficients.

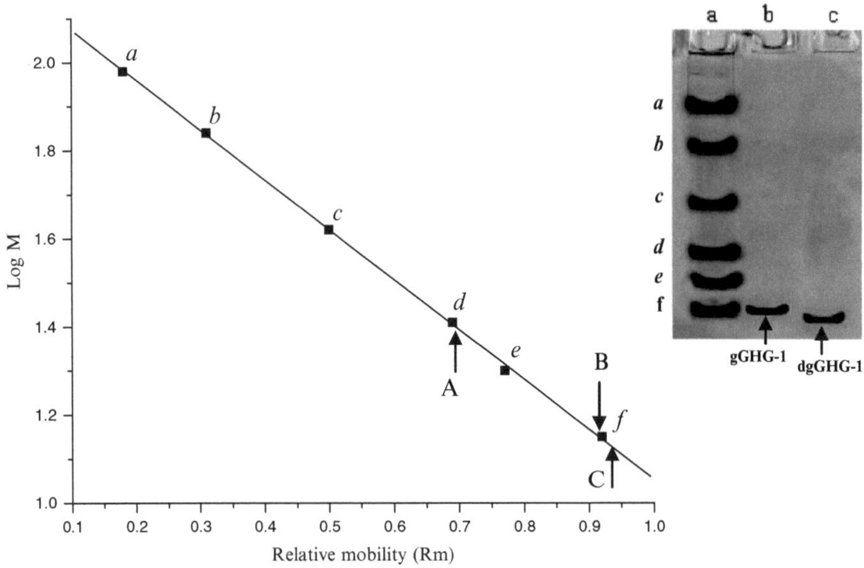

Fig. 10.1 Molecular weight determination using SDS-PAGE: Electrophoresis of the gGHG-1 and dgGHG-1 was performed on 12.5 % acrylamide gel in reducing condition. The relative mobilities (Rm) were plotted against their molecular weight using least square analysis of the data. The molecular weight markers used were: (**a**) Phosphorylase b (97.4 kDa), (**b**) Bovine Serum Albumin (68 kDa), (**c**) Ovalbumin (45 kDa), (**d**) Carbonic Anhydrase (29 kDa), (**e**) Soybean Trypsin Inhibitor (20.1 kDa), (**f**) Lysozyme (14.4 kDa). The elution position of dimer, monomer, and deglycosylated GHG-1 are indicated by *arrows A*, *B*, and *C*, respectively. *Inset* shows the SDS-PAGE of gGHG-1 and dgGHG-1: *Lane* "a" contains molecular weight markers (35 μg) in descending order of their molecular weight, *lane* "b" contains gGHG-1 (35 μg), and *lane* "c" contains dgGHG-1 (35 μg). The gels were stained with CBB R-250 dye

Table 10.1 Comparative physicochemical properties of gGHG-1 and dgGHG-1

Properties	gGHG-1	dgGHG-1
Percent glycosylation	4.25	0.00
Molecular weight (monomer) (kDa)	14.5	14.1
Molecular weight (dimer) (kDa)	29.0	28.2
Stokes radius (monomer) (Å)	17.5	16.4
Stokes radius (dimer) (Å)	26.0	24.9
Diffusion coefficient (monomer) (cm^2/s)	12.7×10^{-15}	13.86×10^{-15}
Diffusion coefficient (dimer) (cm^2/s)	8.57×10^{-15}	9.71×10^{-15}
Temperature optima (°C)	42	39
pH stability range	5.5–9.5	6.5–8.5

Previous studies have shown that the presence of carbohydrate residues in the galectin backbone has significant role in thermal melting of proteins [27–29]. The effect of glycosylation on thermal stability of the purified galectin was therefore curiously investigated. As expected, higher thermal stability of gGHG-1 (100 % activity up to 42 °C) was observed compared with dgGHG-1 (100 % activity up to 39 °C) (Table 10.1). This finding suggests that presence of carbohydrate chains can improve thermal stability of galectins and solubility of the unfolded protein intermediates. The increased thermal stability often correlates with enhanced storage stability of proteins [27].

When exposed to extremes of pH, proteins lose their structure by commotion of charge–charge interactions as well as internal electrostatic forces [30]. Available literature has shown the essentiality of glycosylation in maintaining the conformational stability of proteins against pH denaturation [6]. The effect of glycosylation on pH stability of purified galectin was therefore intriguingly investigated, and as anticipated, the pH stability range of dgGHG-1 (6.5–8.5) was found to be lesser than the pH stability range of gGHG-1 (5.5–9.5) (Table 10.1). Both gGHG-1 and dgGHG-1 forms exhibited highest activity at pH 7.4; nevertheless, a diminished activity profile was observed for dgGHG-1. This finding further establishes the noteworthy role of carbohydrate moiety in guarding proteins from extreme pH conditions, attributed to an increase in the internal electrostatic interactions of the protein pertaining to protein glycosylation [30]. All the physicochemical properties of gGHG-1 and dgGHG-1 discussed till now are comparatively summarized in Table 10.1.

Glycosylation has been reported to augment the conformational stability of proteins against chemically stimulated denaturation [30, 31]. The presence of multiple reactive chemical functionalities in the amino acids' side chains of proteins makes them particularly sensitive to several chemical degradation processes [30]. The preferential action of denaturants, thiol-blocking reagents, and detergents on the inhibition of both gGHG-1 and dgGHG-1 proteins' activity remained unchanged. Amongst thiol-blocking reagents, *p*HMB (maximally hydrophobic) inhibited the activities of both gGHG-1 and dgGHG-1 most readily (Fig. 10.2), trailed by NEM and iodoacetamide (least hydrophobic). For denaturants, urea most readily abolished the activity of gGHG-1 and dgGHG-1 (Fig. 10.3), followed by thiourea and GdnHCl. Between detergents, SDS most readily inhibited the activities of gGHG-1 and dgGHG-1 (Fig. 10.4), trailed by Triton X-100 and Tween 20. Nevertheless, deglycosylated forms exhibited reduced activity profile compared to native galectins, consistent with earlier reports showing the role of glycosylation in imparting resistance to denaturant-mediated inactivation of galectins [6, 32]. These findings advocate that the mechanism responsible for glycosylation-mediated augmentation of the chemical denaturation stability of proteins can be attributed to increased strength of hydrophobic interactions and hydrogen bonding. While augmented strength of hydrophobic interaction can be elucidated by the enhanced stiffness of the glycosylated protein core and superior structural firmness [33], improved strength of hydrogen bonding can be explicated by the abridged water dielectric screening [34].

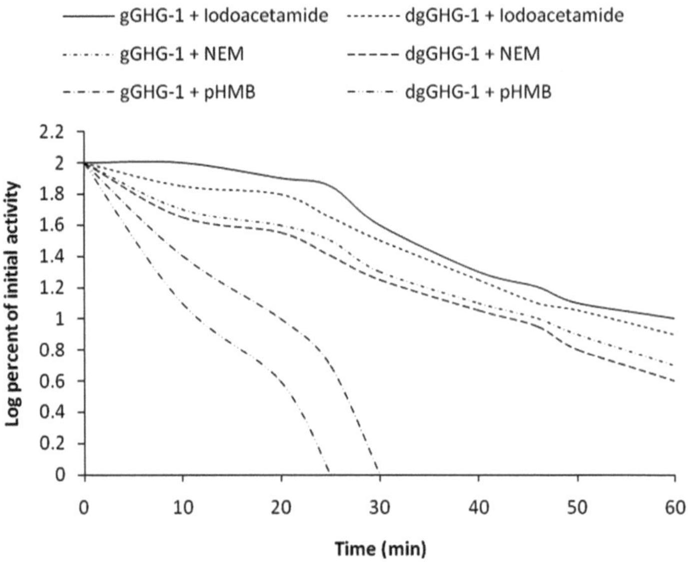

Fig. 10.2 Deglycosylation-mediated inflection in reaction pattern of thiol-blocking agents: The effect of thiol-blocking agents monitored by incubating gGHG-1 and dgGHG-1 (100 μg/ml each in PBS "B") with 70 mM each of iodoacetamide, *p*-hydroxymercuribenzoate (*p*HMB), and N-ethylmaleimide (NEM) at room temperature. After assigned times, all the samples were checked for hemagglutination activity

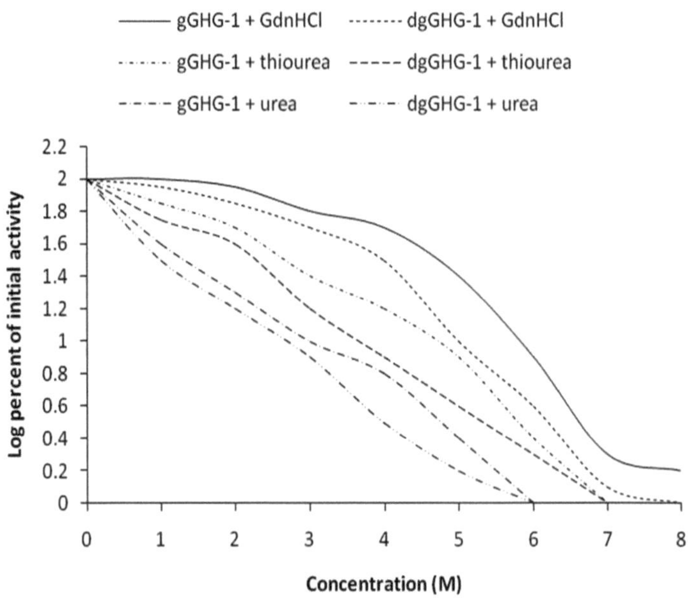

Fig. 10.3 Deglycosylation-mediated inflection in reaction pattern of denaturants: The effect of various denaturants was determined by incubating gGHG-1 and dgGHG-1 (100 μg/ml each in PBS "B") with increasing concentrations (0–8 M) of urea, guanidine HCl (GdnHCl), and thiourea at 37 °C for 1 h. All the samples were then monitored for hemagglutination activity

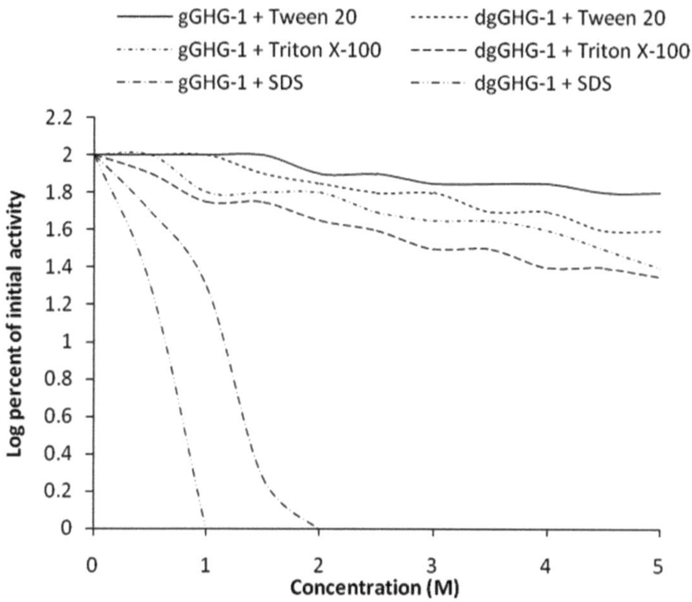

Fig. 10.4 Deglycosylation-mediated inflection in reaction pattern of detergents: The effect of different detergents was monitored by incubating gGHG-1 and dgGHG-1 (100 μg/ml each in PBS "B") with increasing concentrations (1–5 M) of SDS, Tween-20, and Triton X-100 at 37 °C for 1 h. The samples were then observed for hemagglutination activity

The oxidative susceptibility for glycosylated and deglycosylated erythropoietin revealed that glycosylation diminished the protein inactivation as well as the tryptophan oxidation rates, thus highlighting the role of glycosylation in protecting the protein structure from damage by active oxygen species [35, 36]. To further support this finding, both gGHG-1 and dgGHG-1 were oxidized with hydrogen peroxide (H_2O_2), to possibly shed some light on the mechanisms of glycosylation-mediated protein stabilization and whether this stabilizing effect is glycan specific or nonspecific. The UV spectra of gGHG-1 and dgGHG-1 showed maxima at 280 nm, with slightly higher absorbance for dgGHG-1 (Fig. 10.5). Oxidation of gGHG-1 in the presence of 5 mM H_2O_2 (without β-ME) resulted in the shift of the absorption peak to 250 nm, possibly due to oxidation of Trp residue to oxindole moiety [37, 38]. Although the dgGHG-1 exhibited a reduced activity profile pre- and post-oxidation, the effect of oxidation was much more articulated on dgGHG-1. This finding further establishes the protective role of protein glycosylation against oxidant-mediated inactivation. These oxidation events can be mainly attributed to the production of active oxygen-based radicals in protein formulations due to the combination of trace amounts of transition metals, atmospheric oxygen, and exposure to UV light. So far, erythropoietin was the only protein reported to have its bioactivity influenced by oxidation and where glycosylation ameliorates the chemical instability [30, 39]. Our finding thus adds another protein, GHG-1, to this exclusive list. The loss of GHG-1 bioactivity can be easily correlated with the degrees of tryptophan oxidation under oxidizing conditions [30, 40].

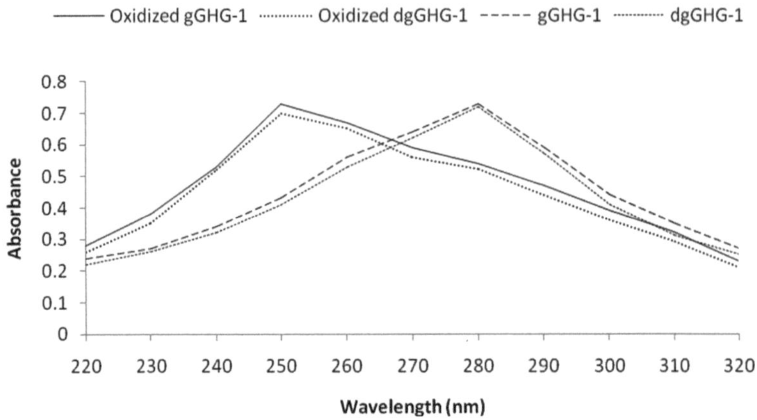

Fig. 10.5 *UV spectra:* The UV spectra of native gGHG-1 and dgGHG-1 (100 μg/ml each in PBS "B" containing β-ME) and oxidized gGHG-1 and dgGHG-1 (100 μg/ml in PBS "B" containing 5 mM H_2O_2 in absence of β-ME) were measured in the wavelength range of 220–320 nm

Fluorescence spectroscopic studies were performed to follow expected deviation in the protein folding post-deglycosylation. The fluorescence emission maxima of gGHG-1 and dgGHG-1 (Fig. 10.6) remained unchanged. Nevertheless, dgGHG-1 demonstrated elevated fluorescence intensity, which can be attributed to the elimination of carbohydrate moieties allowing enhanced exposure of the fluorophore [41]. When exposed to escalating concentrations of urea, *p*HMB, and SDS, the dgGHG-1 exhibited uniformly elevated fluorescence intensity and red shift, thus advocating the significant role of glycan residues (due to steric hindrances) in guarding the Trp residues from more polar microenvironment [42, 43]. Deglycosylation thus resulted in pronounced change in the protein structure, possibly mediated by reorientation of the Trp residues, thus highlighting the importance of glycosylation in stabilizing the protein structure against unfolding.

CD and FTIR spectra of proteins are extremely sensitive to secondary structure, with CD signal being strong for α-helix and weak for β-sheets [44–46]. CD and FTIR spectroscopic studies were therefore carried out to get an insight into the possible variation in the protein secondary structure post-deglycosylation. The far UV-CD spectra (Fig. 10.7) of gGHG-1 as well as dgGHG-1 displayed a negative band near 215 nm and a positive band near 198 nm, and the near UV-CD spectra (Fig. 10.8) of gGHG-1 as well as dgGHG-1 displayed a negative band near 275 nm and a positive band near 260 nm. The FTIR spectra (Fig. 10.9) of gGHG-1 as well as dgGHG-1 displayed an average frequency of near 1,629 cm^{-1} with a minimum of 1,615 cm^{-1} and a maximum of 1,637 cm^{-1}. These findings are typical of β-sheet proteins and characteristic feature of Gal-1, thus further validating our earlier finding depicting inclusion of GHG-1 in galectin-1 family [2, 4, 47]. However, the dgGHG-1 and dgGHG-1 exposed to biologically active reagents (urea, *p*HMB, and SDS) displayed CD and FTIR spectra characteristic for a protein of which all secondary structure is lost, further highlighting the astounding significance of protein glycosylation in protection against damage and denaturation by these chemicals.

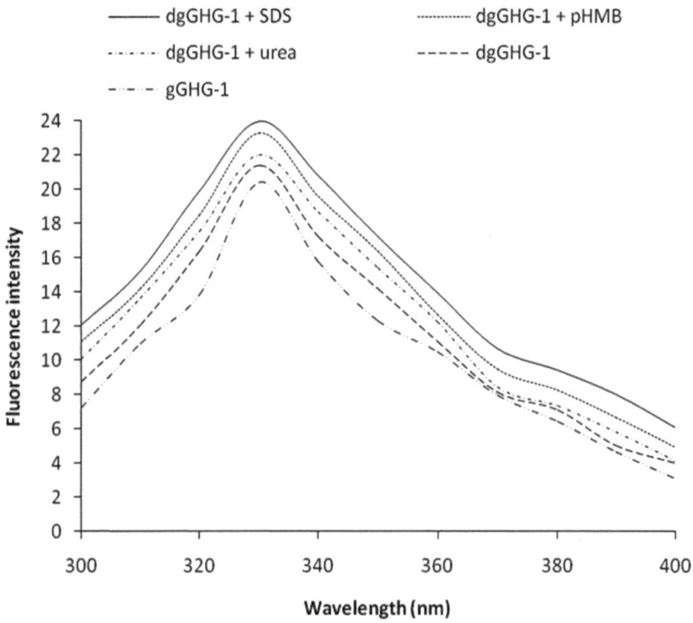

Fig. 10.6 *Fluorescence spectra:* Intrinsic fluorescence of gGHG-1 and dgGHG-1 (100 μg/ml each in PBS "B") was measured by selectively irradiating the protein samples by an excitation wavelength of 280 nm with 10 nm band pass and the emission spectra were measured in the wavelength range of 300–400 nm. Variation in the intrinsic fluorescence of dgGHG-1 in the presence of 5 mM H_2O_2, 8 M urea, 70 mM pHMB, and 5 M SDS was also monitored

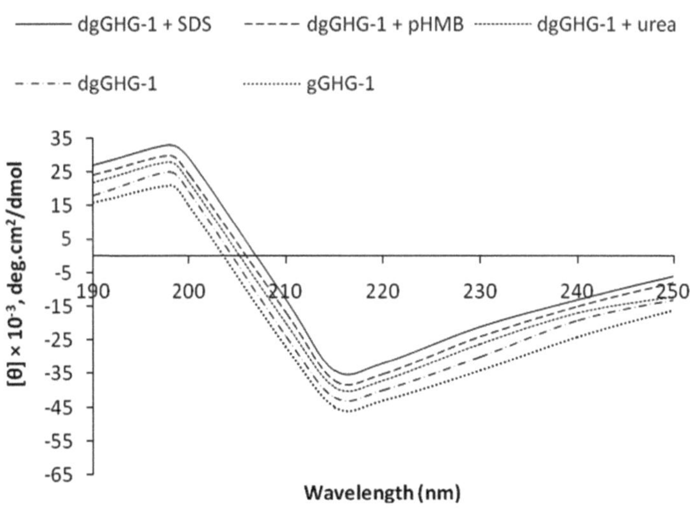

Fig. 10.7 *Far UV-CD spectra:* The far UV-CD spectra of gGHG-1, dgGHG-1, and dgGHG-1 in the presence of 8 M urea, 70 mM pHMB, and 5 M SDS (1 mg/ml each) in PBS "B" were recorded between 190 and 250 nm using spectral bandwidth of 2 nm, scanning speed of 20 nm/min using 0.1 cm path length, and response time of 1 s

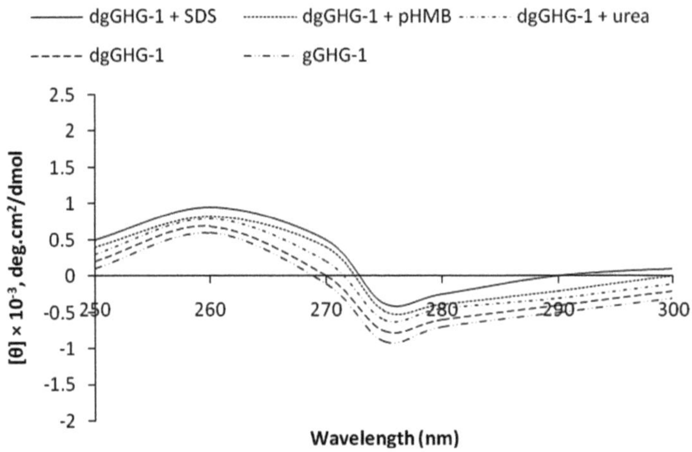

Fig. 10.8 *Near UV-CD spectra:* The near UV-CD spectra of gGHG-1, dgGHG-1, and dgGHG-1 in the presence of 8 M urea, 70 mM pHMB, and 5 M SDS (2 mg/ml each) in PBS "B" were recorded between 250 and 300 nm using spectral bandwidth of 1 nm, scanning speed of 20 nm/min using 0.1 cm path length, and response time of 1 s

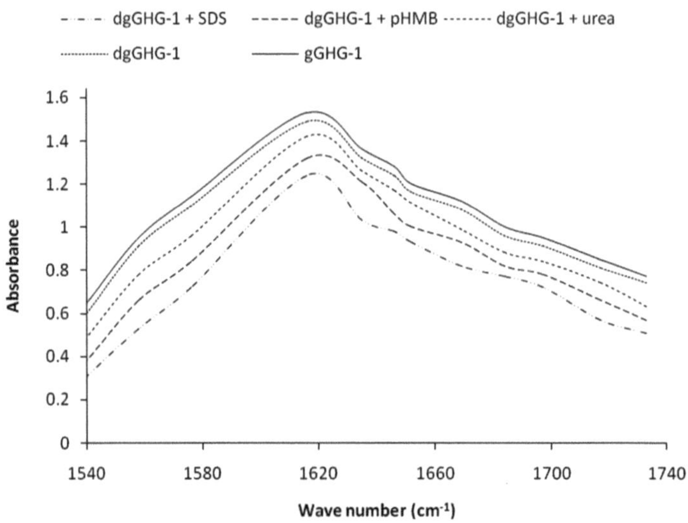

Fig. 10.9 *FTIR spectra:* The FTIR spectra of gGHG-1, dgGHG-1, and dgGHG-1 in the presence of 8 M urea, 70 mM pHMB, and 5 M SDS (150 μg/ml each) in PBS "B" were recorded with a resolution of 3 cm^{-1} and 96 scans

Conclusions

The finding that goat heart galectin-1 is considerably glycosylated formed the basis of the present study. In this study, we for the first time report an extensive comparative physicochemical characterization of glycosylated and deglycosylated forms of GHG-1. The observation that GHG-1 dimerization is not perturbed by deglycosylation suggests that protein glycosylation has apparently no role in the dimerization of protein subunits. The susceptibility of deglycosylated protein to inactivation at extremes of pH and higher temperature indicates the protective role of glycosylation. Enhanced vulnerability of deglycosylated protein towards oxidant-mediated inactivation establishes the protective role of glycosylation, and strengthens our hypothesis "*oxidative inactivation might serve as an important parameter for the measurement of the effect of oxidative assault on the lectin expression under various pathological conditions*" [2], with an additional element of glycosylation-induced galectin stability. Various spectroscopic analyses further highlight the astounding significance of protein glycosylation in protection against damage and denaturation by biologically active chemicals. The present detailed study thus adds an important dimension of the significance of protein expression and altered glycosylation in causation, progression, and possible therapeutics of various cardiovascular and neurological disorders.

Acknowledgements The authors are grateful to Aligarh Muslim University (Aligarh, India) for the facilities. Thanks are due to Deanship of Scientific Research (DSR) and King Fahd Medical Research Center (KFMRC), King Abdulaziz University (Jeddah, Saudi Arabia) for other facilities.

Conflict of interest: Declared none.

References

1. Bacigalupo ML, Manzi M, Rabinovich GA, Troncoso MF (2013) Hierarchical and selective roles of galectins in hepatocarcinogenesis, liver fibrosis and inflammation of hepatocellular carcinoma. World J Gastroenterol 19(47):8831–8849. doi:10.3748/wjg.v19.i47.8831
2. Ashraf GM, Rizvi S, Naqvi S, Suhail N, Bilal N, Hasan S, Tabish M, Banu N (2010) Purification, characterization, structural analysis and protein chemistry of a buffalo heart galectin-1. Amino Acids 39(5):1321–1332. doi:10.1007/s00726-010-0574-7
3. Hasan SS, Ashraf GM, Banu N (2007) Galectins – potential targets for cancer therapy. Cancer Lett 253(1):25–33. doi:10.1016/j.canlet.2006.11.030
4. Camby I, Mercier ML, Lefranc F, Kiss R (2006) Galectin-1: a small protein with major functions. Glycobiology 16(11):137R–157R. doi:10.1093/glycob/cwl025
5. Barondes SH, Cooper DN, Gitt MA, Leffler H (1994) Galectins. Structure and function of a large family of animal lectins. J Biol Chem 269(33):20807–20810

6. Ashraf GM, Bilal N, Suhail N, Hasan S, Banu N (2010) Glycosylation of purified buffalo heart galectin-1 plays crucial role in maintaining its structural and functional integrity. Biochemistry 75(12):1450–1457
7. Bardosi A, Bardosi L, Hendrys M, Wosgien B, Gabius HJ (1990) Spatial differences of endogenous lectin expression within the cellular organization of the human heart: a glycohistochemical, immunohistochemical, and glycobiochemical study. Am J Anat 188(4):409–418. doi:10.1002/aja.1001880409
8. Ashraf GM, Banu N, Ahmad A, Singh LP, Kumar R (2011) Purification, characterization, sequencing and biological chemistry of galectin-1 purified from Capra hircus (goat) heart. Protein J 30(1):39–51. doi:10.1007/s10930-010-9300-2
9. Shin T, Carrillo-Salinas FJ, Feliu Martinez A, Mecha M, Guaza C (2013) Immunohistochemical localization of galectin-3 in the brain with Theiler's murine encephalomyelitis virus (DA strain) infection. Korean J Vet Res 53(3):159–162
10. Stillman BN, Mischel PS, Baum LG (2005) New roles for galectins in brain tumors – from prognostic markers to therapeutic targets. Brain Pathol 15(2):124–132. doi:10.1111/j.1750-3639.2005.tb00507.x
11. Qiao X, Tian J, Chen P, Wang C, Ni J, Liu Q (2013) Galectin-1 is an interactive protein of selenoprotein M in the brain. Int J Mol Sci 14(11):22233–22245. doi:10.3390/ijms141122233
12. Starossom Sarah C, Mascanfroni Ivan D, Imitola J, Cao L, Raddassi K, Hernandez Silvia F, Bassil R, Croci Diego O, Cerliani Juan P, Delacour D, Wang Y, Elyaman W, Khoury Samia J, Rabinovich Gabriel A (2012) Galectin-1 deactivates classically activated microglia and protects from inflammation-induced neurodegeneration. Immunity 37(2):249–263. doi:10.1016/j.immuni.2012.05.023
13. Jin J-K, Na Y-J, Song J-H, Joo H-G, Kim S, Kim J-I, Choi E-K, Carp RI, Kim Y-S, Shin T (2007) Galectin-3 expression is correlated with abnormal prion protein accumulation in murine scrapie. Neurosci Lett 420(2):138–143. doi:10.1016/j.neulet.2007.04.069
14. Wang X, Zhang S, Lin F, Chu W, Yue S (2013) Elevated galectin-3 levels in the serum of patients with Alzheimer's disease. Am J Alzheimer's Dis Other Demen. doi:10.1177/1533317513495107
15. Lerman BJ, Hoffman EP, Sutherland ML, Bouri K, Hsu DK, Liu F-T, Rothstein JD, Knoblach SM (2012) Deletion of galectin-3 exacerbates microglial activation and accelerates disease progression and demise in a SOD1(G93A) mouse model of amyotrophic lateral sclerosis. Brain Behav 2(5):563–575. doi:10.1002/brb3.75
16. Lis H, Sharon N (1993) Protein glycosylation. Structural and functional aspects. Eur J Biochem 218(1):1–27
17. Komatsu S, Yamada E, Furukawa K (2009) Cold stress changes the concanavalin A-positive glycosylation pattern of proteins expressed in the basal parts of rice leaf sheaths. Amino Acids 36(1):115–123. doi:10.1007/s00726-008-0039-4
18. Hughes RC (1999) Secretion of the galectin family of mammalian carbohydrate-binding proteins. Biochim Biophys Acta 1473(1):172–185
19. Erickson HP (2009) Size and shape of protein molecules at the nanometer level determined by sedimentation, gel filtration, and electron microscopy. Biological Procedures Online 11:32–51. doi:10.1007/s12575-009-9008-x
20. Lowry OH, Rosebrough NJ, Farr AL, Randall RJ (1951) Protein measurement with the Folin phenol reagent. J Biol Chem 193(1):265–275
21. Raz A, Lotan R (1981) Lectin-like activities associated with human and murine neoplastic cells. Cancer Res 41(9 Pt 1):3642–3647
22. DuBois M, Gilles KA, Hamilton JK, Rebers PA, Smith F (1956) Colorimetric method for determination of sugars and related substances. Anal Chem 28(3):350–356. doi:10.1021/ac60111a017
23. Yamabayashi S (1987) Periodic acid – Schiff – Alcian blue: a method for the differential staining of glycoproteins. Histochem J 19(10–11):565–571. doi:10.1007/bf01687364

24. Rasheedi S, Haq SK, Khan RH (2003) Guanidine hydrochloride denaturation of glycosylated and deglycosylated stem bromelain. Biochemistry 68(10):1097–1100
25. Weber K, Osborn M (1969) The reliability of molecular weight determinations by dodecyl sulfate-polyacrylamide gel electrophoresis. J Biol Chem 244(16):4406–4412
26. Gomord V, Faye L (2004) Posttranslational modification of therapeutic proteins in plants. Curr Opin Plant Biol 7(2):171–181. doi:10.1016/j.pbi.2004.01.015
27. Arnold U, Ulbrich-Hofmann R (1997) Kinetic and thermodynamic thermal stabilities of ribonuclease A and ribonuclease B. Biochemistry 36(8):2166–2172. doi:10.1021/bi962723u
28. DeKoster GT, Robertson AD (1997) Thermodynamics of unfolding for Kazal-type serine protease inhibitors: entropic stabilization of ovomucoid first domain by glycosylation. Biochemistry 36(8):2323–2331. doi:10.1021/bi962580b
29. Watanabe M, Nakamura O, Muramoto K, Ogawa T (2012) Allosteric regulation of the carbohydrate-binding ability of a novel conger eel galectin by D-mannoside. J Biol Chem 287(37):31061–31072. doi:10.1074/jbc.M112.346213
30. Sola RJ, Griebenow KAI (2009) Effects of glycosylation on the stability of protein pharmaceuticals. J Pharm Sci 98(4):1223–1245. doi:10.1002/jps.21504
31. Sarkar A, Wintrode PL (2011) Effects of glycosylation on the stability and flexibility of a metastable protein: the human serpin? 1-antitrypsin. Int J Mass Spectrom 302(1–3):69–75. doi:10.1016/j.ijms.2010.08.003
32. Solá RJ, Rodríguez-Martínez JA, Griebenow K (2007) Modulation of protein biophysical properties by chemical glycosylation: biochemical insights and biomedical implications. Cell Mol Life Sci 64(16):2133–2152. doi:10.1007/s00018-007-6551-y
33. Abdelkebir K, Gaudière F, Morin-Grognet S, Coquerel G, Atmani H, Labat B, Ladam G (2011) Protein-triggered instant disassembly of biomimetic layer-by-layer films. Langmuir 27 (23):14370–14379. doi:10.1021/la2033109
34. Barrozo A, Borstnar R, Marloie G, Kamerlin SCL (2012) Computational protein engineering: bridging the gap between rational design and laboratory evolution. Int J Mol Sci 13 (10):12428–12460. doi:10.3390/ijms131012428
35. Flora SJS (2009) Structural, chemical and biological aspects of antioxidants for strategies against metal and metalloid exposure. Oxid Med Cell Longev 2(4):191–206
36. Martínek V, Sklenář J, Dračínský M, Šulc M, Hofbauerová K, Bezouška K, Frei E, Stiborová M (2010) Glycosylation protects proteins against free radicals generated from toxic xenobiotics. Toxicol Sci 117(2):359–374. doi:10.1093/toxsci/kfq206
37. Levi G, Teichberg VI (1981) Isolation and physicochemical characterization of electrolectin, a beta-D-galactoside binding lectin from the electric organ of Electrophorus electricus. J Biol Chem 256(11):5735–5740
38. Vincenzetti S, Cambi A, Maury G, Bertorelle F, Gaubert G, Neuhard J, Natalini P, Salvatori D, Sanctis GD, Vita A (2000) Possible role of two phenylalanine residues in the active site of human cytidine deaminase. Protein Eng 13(11):791–799. doi:10.1093/protein/13.11.791
39. Plugin glycosylation literature survey ricky connolly 091211-1 (trans: Ricky C)
40. Kim YH, Berry AH, Spencer DS, Stites WE (2001) Comparing the effect on protein stability of methionine oxidation versus mutagenesis: steps toward engineering oxidative resistance in proteins. Protein Eng 14(5):343–347. doi:10.1093/protein/14.5.343
41. Ketko AK, Lin C, Moore BB, LeVine AM (2013) Surfactant protein A binds flagellin enhancing phagocytosis and IL-1? production. PLoS One 8(12). doi: 10.1371/journal.pone.0082680
42. Torrent J, Alvarez-Martinez MT, Liautard J-P, Balny C, Lange R (2005) The role of the 132–160 region in prion protein conformational transitions. Protein Sci 14(4):956–967. doi:10.1110/ps.04989405
43. Fatima A, Husain Q (2007) A role of glycosyl moieties in the stabilization of bitter gourd (Momordica charantia) peroxidase. Int J Biol Macromol 41(1):56–63. doi:10.1016/j.ijbiomac.2006.12.007

44. Chirgadze YN, Nevskaya NA (1976) Infrared spectra and resonance interaction of amide-I vibration of the antiparallel-chain pleated sheet. Biopolymers 15(4):607–625. doi:10.1002/bip.1976.360150402
45. Huerta-Viga A, Woutersen S (2013) Protein denaturation with guanidinium: A 2D-IR study. J Phys Chem Lett 4(20):3397–3401. doi:10.1021/jz401754b
46. Johnson WC (1988) Secondary structure of proteins through circular dichroism spectroscopy. Annu Rev Biophys Biophys Chem 17(1):145–166. doi:10.1146/annurev.bb.17.060188.001045
47. Mandal P, Molla AR, Mandal DK (2013) Denaturation of bovine spleen galectin-1in guanidine hydrochloride and fluoroalcohols: Structural characterization and implications for protein folding. J Biochem. doi:10.1093/jb/mvt084

Chapter 11
Analytic Considerations and Axiomatic Approaches to the Concept Cell Death and Cell Survival Functions in Biology and Cancer Treatment

Ioannis Gkigkitzis*, **Ioannis Haranas***, **and Carlos Austerlitz**

Abstract This study contains a discussion on the connection between current mathematical and biological modeling systems in response to the main research need for the development of a new mathematical theory for study of cell survival after medical treatment and cell biological behavior in general. This is a discussion of suggested future research directions and relations with interdisciplinary science. In an effort to establish the foundations for a possible framework that may be adopted to study and analyze the process of cell survival during treatment, we investigate the organic connection among an axiomatic system foundation, a predator–prey rate equation, and information theoretic signal processing. A new set theoretic approach is also introduced through the definition of cell survival units or cell survival units indicating the use of "proper classes" according to the Zermelo–Fraenkel set theory and the axiom of choice, as the mathematics appropriate for the development of biological theory of cell survival.

11.1 Introduction

The goal of the study of cell decision processes according to the National Institute of General Medical Sciences of NIH is to understand how molecular signaling networks regulate cell life–death decisions in tissues and to apply this understanding to

*Author contributed equally with all other contributors.

I. Gkigkitzis*
Departments of Mathematics and Biomedical Physics, East Carolina University,
124 Austin Building, East Fifth Street, Greenville, NC 27858-4353, USA
e-mail: gkigkitzisi@ecu.edu

I. Haranas* (✉)
Physics and Computer Science, Wilfrid Laurier University, Science Building, Room N2078
75 University Ave. W, Waterloo, ON, N2L 3C5, Canada
e-mail: yiannis.haranas@gmail.com

C. Austerlitz
Department of Computer Science, SIU Southern Illinois University,
Carbondale, Illinois 62902, USA

P. Vlamos, A. Alexiou (eds.), *GeNeDis 2014*, Advances in Experimental Medicine and Biology 822, DOI 10.1007/978-3-319-08927-0_11

the design and use of therapeutic drugs[1]. Research in this direction aims at proper application of mathematical modeling and experimental techniques that will yield quantitative understanding of cellular decision-making processes and their correlations with cellular biochemistry. The validity of mathematical models can only be established by rigorous analysis and comparison with experimental results.

The 2012 Noble Prize in physiology or medicine went to John Gurdon and Shinya Yamanaka for turning mature cells into stem cells. Yamanaka shook the world just 6 years ago in a Cell paper [45] that showed how to reprogram adult fibroblast cells into pluripotent stem cells (iPS cells) by simply inducing four genes—Oct 3/4, Sox2, c-Myc, and Klf4. Although he may not frame it this way, Yamanaka arrived at these four genes by applying a simple theorem of formal logic, which is that a set of AND conditions is equivalent to negations of OR conditions. For example, the statement A AND B is True is the same as Not A OR Not B is False. In formal logic notation you would write $A \wedge B = \neg(\neg A \vee \neg B)$.

However, the empirical character of biology and biological processes gives evidence to the absence of a global meaningful logical theory of all biological phenomena. Any conceptual innovation cannot in practice be carried out unless a new formal scheme has in the first place been made available. In the transitions from physics (or mathematics) to biology the innovations are logical rather than mathematical. In order to carry through a new program that would accommodate and also remove any contradictions between the laws of physics and the regularities of biology, one has to recognize what are the independent formal elements of scientific description, on a par with the familiar laws of nature. This would be a decisive novel point which would define a program. At this point we can do no better than to recommend the foundation for such a program to the consideration of our colleagues. The current study includes the introduction of (and points towards) an axiomatic foundation of biological behavior of cells, according to the current line of research in the field that will assess various experimental data.

Mathematical modeling is essential to clearly reveal and understand the fundamental processes underlying the properties of living biophysical systems. It allows the abstraction, idealization, and description of major characteristics depending upon the interest of the research. For example, intracellular molecular interactions can be studied with the purpose of extracting useful conclusions, by using computational methods. The mathematical formulation of biochemical reactions of enzyme catalysis [9, 51, 53] has been used as a standard reference in molecular dynamics with a number of applications beyond enzymatic physics. Another example comes from survival analysis. Survival Analysis is the branch of Statistics that studies death in biological organisms and failure in mechanical systems. The theory is based on the axiom that death (failure) occurs once for each subject although recurring event or repeated event models can be adapted by assuming partial failure, e.g., heart attack. The object function to be studied in a survival analysis is the survival function [21].

[1] Center for Cell Decision Processes (MIT), National Centers for System Biology. http://www.systemscenters.org/centers/center-for-cell-decision-processes/.

A question of major importance is if a cell is damaged, does it repair itself or decide to self-destruct and how does the cell make that decision? Existing models develop multi-scale approaches of how single-molecule events integrate networks to determine cell fate. Indeed, cancer in particular is the result of cell decision-making and deserves special focus. Other activities such as cell maintenance, replication, repair, recombination, programmed cell death—apoptosis, etc. need to be investigated. Cellular decision-making is defined as the process whereby cells assume different, functionally important and heritable fates without an associated genetic or environmental difference [5]. These questions are directly related to cancer and cancer therapy. One defining feature of cancer is the rapid creation of abnormal cells that grow beyond their usual boundaries, and which can then invade adjoining parts of the body and spread to other organs. This process is referred to as metastasis. Metastases are the major cause of death from cancer. Cancers are primarily an environmental disease with 90–95 % of cases attributed to environmental factors and 5–10 % due to genetics [2], although it is nearly impossible to prove what caused a cancer in any individual, because most cancers have multiple possible causes. Medical tests commonly used include blood tests, X-rays, CT scans, and endoscopy. The major management options for cancer are surgery, chemotherapy, radiation, and palliative care.

11.1.1 Continuous vs. Discrete Modeling for Cell Survival: An Intuitive Example

One of the most often asked questions relates to whether a continuous or discrete modeling tool should be used to model the behavior of cell model under a certain treatment, for further simulations and analysis. Of course the answer depends on several aspects of the cell model to be considered and it depends even more on the perspective one wishes to take with regard to the system. In [23], the analog of the bathtub that is filling with discrete drops of water, which we perceive as a continuous flow, is considered. The claim is that if someone watches as water flows into a bathtub, once he turns the faucet on, it appears to be a rather continuous filling process. Thus what he perceives is a continuous process as he watches the bathtub fill little by little over time. If he were instead to turn the faucet on and then leave the room, and provided he returned before the tub overflowed, on his return he would notice that that bathtub was full. So over a period of some number of minutes what he perceives is a discrete change in the bathtub water level, from empty to full. And if he were riding on a water molecule experiencing time in nanoseconds this would all appear a very continuous process [23]. We extend this observation and consider an observer, initially in the bathtub. The level changes when an observer gets out of the bathtub, and the changed level appears as a discrete step, rather than a continuous curve. The changed level and its stepped appearance do not negate the continuous nature of system [23]. Adopting this logic, we replace the bathtub by the biochemistry of a cell, the drops of water by discrete quantized units of cell survival probability (biochemical survival units as defined below, survival and life units will be used interchangeably), and the total water level in the bathtub by the continuous cell survival probability (biochemical life of the cell), which now is the sum of discrete quantized cell survival probabilities. This results in a duality in the perception of the nature of probability of survival of the cell. This may allow the interpretation of survival probability as a discrete population, whose dynamics can be modeled using continuous equations; in particular predator–prey equations where the prey is the life of the cell and the predator is the cell death effectors.

11.2 Life Units Duality

Life Units Duality refers to the idea that the life a cell can be discretized (quantized) in quanta of life which are assumed here as the basic units of life in every cell. New cells are produced by existing cells, and therefore the termination of a cell does not allow to assign any morphological or biochemical characteristics to the life of the cell itself, since these can only be considered as the manifestations of the monitoring, interaction, and response of the cell, as a biochemical unit undividedly united to cellular life ("life units"), to the extracellular environment. Cellular life is a set of life units, where each cellular life unit contains the whole complete life of the cell in itself, therefore allowing the cell to repair itself after any loss of life units due to the attack of cell death inducers or other factors. In analogy to many-particle physics one replaces the actual cell life by a cellular life density. This life density follows certain axioms:

(a) Each cell's life is made up of finite number of survival units.
(b) All have the same life or number of life units and cells of the same type manifest these life units through similar biochemical patterns.
(c) Survival units cannot be described biochemically, although they result in different patterns of the cellular biochemistry.
(d) The activity of an organism depends on the total activity of interdependent life units.
(e) Each cellular life unit contains the whole complete life of the cell in itself (Figs. 11.1 and 11.2).

11.2.1 Russell's Paradox and Information

The suggestion that each cellular life unit contains the whole complete life of the cell in itself (and therefore all survival units, including itself) is related to Mathematics and its generally accepted axioms of set theory. A set is a collection of objects, or elements. Sets are defined by the unique properties of their elements and sets and elements may not be mentioned simultaneously, since sets are determined

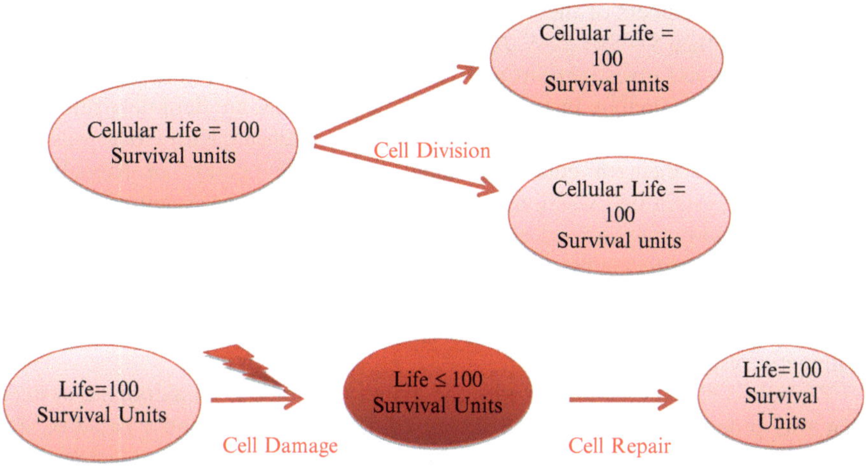

Fig. 11.1 A schematic representation of the idea of cell survival units modeling. Cell life as a sum of survival units (quanta) and a life quantum (a "survival probability quantum") contains life in an embryonic form that might or might not develop to a 100 % probability of survival depending on the state of the cell (e.g., recovery after photo-chemotherapeutic treatment, stress, hyperthermia, etc.)

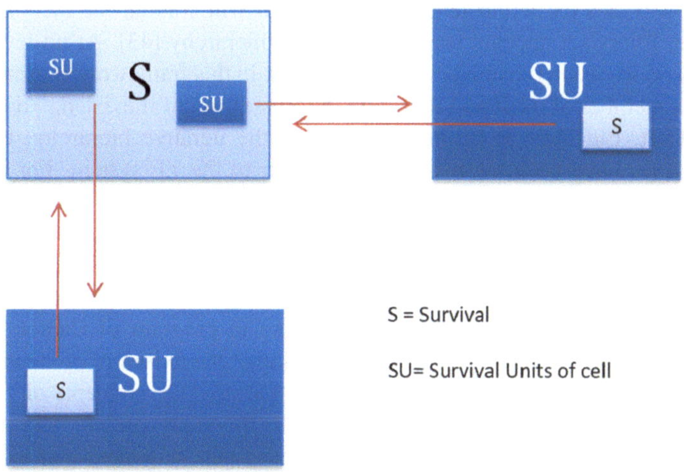

Fig. 11.2 A schematic representation of the idea of cell survival units modeling. Cell life as a sum of survival units (quanta) and a life quantum (a "survival probability quantum") contains life in an embryonic form that might or might not develop to a 100 % probability of survival depending on the state of the cell (e.g., recovery after photo-chemotherapeutic treatment, stress, hyperthermia, etc.)

by their elements and therefore one notion has no meaning without the other, according to Peano [16, 22, 52]. Bertrand Russell, while working on his "Principia Mathematica" (Principles of Mathematics) in 1903, discovered a paradox that arose from Frege's set theory that leads to a contradiction [29]. It says "the set of all sets which are not members of themselves contains itself." In mathematical terms, let $S = \{x : x \notin x\}$, then $S \in S \Leftrightarrow S \notin S$. Although the precise rules for set formation have been under intense investigations and several different logical systems have been proposed, sets that contain themselves as elements, like S, are definitely ruled out, as "abnormal." Based on the work of Russell and Whitehead, Kurt Gödel was able to show that a theorem could be stated within the context of Russell and Whitehead's system that was impossible to prove within that system [47]. Gödel created a paradox that showed a theorem could be true within the framework of *Principia Mathematica* but was also not provable by the rules of Russell's *Principia Mathematica*.[2] Gödel's Incompleteness Theorem states that there are mathematical statements that can never be proved, in any consistent system of axioms such as the arithmetic system.

Zermelo–Fraenkel set theory with the axiom of choice is an axiomatic system that was proposed to formulate a theory of sets without Russell's paradox. ZF embodies to a degree a certain conception of set which is called "combinatorial" or "iterative conception" [43]. The formation of a set starts with some individuals, collected together to form a set. Suppose we start with individuals at the lowest level. At the next level, we form sets of all possible combinations of these individuals. And then we iterate this procedure: at the next level, we form all possible sets of sets and individuals from the first two levels and so on. Given the set of all sets at a particular level, the next level will contain the members of its power set. Every set appears somewhere in the hierarchy [43]. At no level of the hierarchy do we reach the universal set of all sets; in this framework it turns out that no set is a member of itself and therefore the Russell set, if it existed, would be the universal set. But there is no universal set in the iterative hierarchy. However, several issues have been addressed with respect to the ZF system. For example, issues arise in the ZF quantification over sets, and a domain of quantification is needed, but no set in the hierarchy can serve as this domain [43].

To overcome these issues, two alternatives have been suggested. One is the "new foundations"(NF) system, introduced by Quine [41], and the other is the prospects for a set theory with a universal set, according to program of Cantor and Von Neumann. The NF system is based on two axioms: the axiom of "extensionality" and the axiom of a "comprehension schema" that uses the concepts of "stratified formulas." A substantial difficulty appeared in the study of NF. The axiom of choice (AC) is an axiom of set theory equivalent to the statement that the product of a collection of nonempty sets is nonempty. Specker has shown that the axiom of choice fails in NF [44]. This evidence indicated that one should probably follow the alternative of admitting a universal set, with subcollections that are not sets [49]. A semiset is a subclass of a set, and a proper semiset is a subclass of a set that is not itself a set. Semisets are given via properties and predication (the attributing of

[2] Kelly LaFleur, Russell's Paradox, Department of Mathematics University of Nebraska-Lincoln, July 2011.

characteristics to a subject to produce a meaningful statement combining verbal and nominal elements, a propositional function, encyclopedia Britannica) [43].

The need for the distinction between two kinds of collection [43] can be found back in the work of Schroder and Cantor: "If we start from the notion of a definite multiplicity of things, it is necessary, as I discovered, to distinguish two kinds of multiplicities (by this I always mean definite multiplicities). For a multiplicity can be such that the assumption that all of its elements 'are together' leads to a contradiction, so that it is impossible to conceive of the multiplicity as a unity, as 'one finished thing.' Such multiplicities I call absolutely infinite or inconsistent multiplicities.... If on the other hand the totality of the elements of a multiplicity can be thought of without contradiction as 'being together,' so that they can be gathered together into 'one thing,' I call it a consistent multiplicity or a 'set.'"[3]

Cantor's conclusions are the ancestors of today's distinction between classes and sets, as they appear in the work of Von Neumann [48]. For von Neumann all sets are classes, but not all classes are sets. And those classes that are not sets—the so-called proper classes—cannot themselves be members [43]. In Von Neumann's axiomatization theory, some major advantages are [43]: There are extensions for the predicates "set," "non-self-membered set," "well-founded set," and "ordinal." There is a well-determined collection of all the ZF sets; and there is a domain for quantification over sets. Further, the Axiom of Choice is provable in von Neumann's system. Several issues, both technical and intuitive, have been reported with respect to this system. A discussion can be found in [43], and here we only mention the consequence of this theory, that the concept of class has no extension (based on the axioms of this system, there is no class of all classes, and therefore the problem has just been pushed back).

An alternative approach involving "extensions" was suggested for the resolution of Russell's paradox. The extension of a predicate is the set of tuples of values that, used as arguments, satisfy the predicate (a truth valued function). Such a set of tuples is a relation. But this has been shown to be a pathological predicate logic case [43] with respect to universal extensions (Russell's paradox pushed back again). Therefore the resolution of this paradox remains unresolved.

11.2.2 When Is the Cell Dead?

In mathematical logic, it is suggested that problems that are essentially the same must be resolved by the same means, and similar paradoxes should be resolved by similar means. This is the principle of uniform solution. Two paradoxes can be thought to be of the same kind when (at a suitable level of abstraction) they share a similar internal structure, or because of external considerations such as the relationships of the paradoxes [11]. The question arises as to the existence of other

[3] Cantor [8] and Schroder [4], cited in van Heijenoort [47], copied from Simmons [43].

paradoxes that are of the same kind with Russell's paradox. Russell focused more on the underlying structure of the paradoxes and saw them all as paradoxes of impredicativity. The "inclosure schema" was proposed by Priest, as a formal schema that can be used to classify paradoxes [40]. Although the schema will not be analyzed in this work, the conclusion is very interesting: Russell's paradox is of one kind with the "sorites" paradox (the paradox of the "heap"). This paradox was introduced by Eubulides of Miletus (fourth century BC), a pupil of Euclid, and appears when one considers a heap of sand, from which grains are removed. Is it still a heap when only one grain remains? If not, when did it change from a heap to a non-heap? These two paradoxes are neighboring paradoxes, and it has been suggested that we should not just consider the internal structure of the paradoxes—although that is undoubtedly important—we also consider the external relationships—the relationships to other nearby paradoxes[4]. The way nearby neighbors (paradoxes of one kind) respond or fail to respond to proposed treatments tells us something about what makes the whole family tick and about their structural similarity.

The question "when is the cell dead?" indicates confusion between cessation of organic coherence and cellular activity. When a cell irrevocably loses its organization, it's dead. The point when it becomes irrevocably damaged is related to the sorites problem. The sorites paradox appears in the conventional definition of amount of substance [50]. The amount of substance n is as a quantity proportional to number of entities N. This implies that n is discrete for small N while n is considered to be continuous at the macroscopic scale, leading to a sorites paradox. A practical criterion has been proposed in [50] for distinguishing between amount of substance and number of entities, that is to resolve this case of the sorites paradox. In this study, the ideal gas equation $PV = nRT$ is derived as a combination of Boyle's law $P \propto 1/V$, of Charles law $V \propto T$, and Avogadro's law $N \propto V$. By substituting the molar gas constant by the Boltzmann constant, a brief analysis of the resulting ideal gas equation $PV = Nk_BT$ for the case $N = 1$ (the well-known "particle in a box" situation) leads to quantization of energy, and therefore quantization of temperature. However, kinetic theory assumes a large number of particles and this brings up the sorites paradox as the question of what is the scale at which we can consider temperature to be a continuous quantity of kinetic theory to be valid. An alternative metrological criterion for large N was proposed [50]:

- Consider a physical quantity that depends on the amount of substance: it will obviously also depend on the number of entities, and its numerical value can be expressed as $f(x)$ where x is the numerical value of N.
- Consider a measurement of that quantity for $N = x$ and for $N = x + 1$. These measurements will be associated with measurement uncertainties.
- If the difference between the two measurement results is significant with respect to the uncertainties in those measurement results, the quantity is considered to be

[4] Colyvan Mark, "Vagueness and Truth", in H. Dyke (ed.), From Truth to Reality: New Essays in Logic and Metaphysics, Routledge, 2009,pp. 29–40.

discrete at $N = x$; if not, the quantity is treated as continuous for that measurement procedure at that scale.

In other words, if there is a significant difference in measurement result by adding a single entity the measurement is a count of number of entities; if there is no significant difference in measurement result on adding a single entity, it is a measurement of amount of substance. Temperature is one of the uncertainty sources in precision dimensional measurement and probability density functions of temperature change are usually derived by mathematical models. Moreover, temperature is the ability of one body to give up heat energy to another body. Temperature measures the concentration of thermal energy in an object in much the same way that density measures the concentration of matter in an object.

The Ehrenfest model of diffusion was originally proposed as a model for dissipation of heat and to explain the second law of thermodynamics. The model is defined by a system of N particles in two containers, with particles independently change container. The stochastic process $X = (X_1, X_2, \ldots)$ of the state of the system is defined by $X(n) = X_n$, the number of particles in one container at time $n \in N$, and the state space is $S = \{0, 1, 2, 3 \ldots, m\}$ where m is the total number of particles. The system evolves according to the transition probability $P(x, x-1) = \frac{x}{m}$, $P(x, x+1) = \frac{m-x}{m}$, $x \in S$. A generalized form of this model is $P(x, x-1) = \lambda x$, $P(x, x+1) = \lambda(m-x)$, where λ is the transition rates. This last Markov process method can be used in dynamic probabilistic systems to make sequential predictions, where the system can be in a finite number of states and the decision-making process involves a choice of several actions in each of those states. Solution techniques for Markov decision problems rely on exact knowledge of the transition rates, which may be difficult or impossible to obtain and therefore current studies focus on the quantification of the range of the uncertainty of the transition rates [20]. Despite this, one can observe that the transition probabilities can be thought of as the measurable "physical quantity" that depends on the amount of substance, the number of entities. These measurements are indeed associated with measurement uncertainties, the source of which is the transition rates. The metrological criterion that was introduced above can be applied here, indicating the presence of the sorites problem for discreteness vs. continuity in the definition of the probability. Moreover, it is known that continuous time Markov processes are used for the formulation of stochastic predator–prey models that are based on within-individual variation [6, 10, 24]. Within-individual variation refers to the fact that no considerations are taken of characteristics of an individual that affect its chance of dying. Death is treated as an intrinsically within-individual phenomenon [10]. For example, chance effects may lead to death of some individuals and these outcomes are likely regardless of the characteristics of the individuals involved. The variation is unique to the individual, but is unpredictable, given any particular characteristics.

11.2.3 Demographic Stochasticity and Internal Quantization

Within-individual variation, used under the name of "demographic stochasticity," has been used in the theory of adaptive dynamics. The theory of adaptive dynamics aims at describing the dynamics of the dominant trait in a population, which is called the "fittest" trait. The main approach is through stochastic or individual centered models which in the limit of large population can be transformed into integro-differential equations or partial differential equations [8, 12, 39]. Stochastic simulations, using a finite size population, involve extinction phenomenon operating through demographic stochasticity (which is another name for the "within-individual variation") which acts drastically on small populations [39]. These simulations involve a unit for minimal survival population size, which corresponds to a single individual. In general though, typical stochastic and deterministic simulations do not fit and give rather different behaviors in terms of branching patterns. It has been observed that the notion of demographic stochasticity does not occur in general in deterministic population models, and an alternative approach has been proposed in order to include a similar notion in these models: the notion of a survival threshold [34], which allows some phenotypical traits of the population to vanish when represented by too few individuals. In particular, through the investigations of simple and standard Lotka Volterra systems that describe the time of the distribution of phenotypic traits in time, it is shown that the inadequacy of deterministic models to handle extinction phenomena through demographic stochasticity can be corrected by the introduction of a survival threshold, leading to a mimicking effect of the extinction probability due to demographic stochasticity in small subpopulations, while hardly influencing the dynamics of large subpopulations [39]. In this framework, the above principle implies (at the extreme) that densities corresponding to less than one individual are undesirable [39], indicating that the link between the continuous (large populations) and the discrete (small subpopulations), between the existence (survival) and the vanishing (extinction—demographic stochasticity), between the deterministic approach (differential equations) and the stochastic approach is correlated with the existence of a survival threshold in the model, originating from the discreteness part of this duality model.

This hybrid approach of survival, as continuous-discrete function with a survival threshold assigned to a population, raises the following question: Is there an internal quantization scheme that relates the continuous models for large populations with survival thresholds to small population discrete models? As mentioned above, the one is in agreement with the other in the appropriate limits, but the presence of the limit involves the external operation of rescaling, which is related according to our previous discussion to the sorites paradox. In particular, the existence of both features, of continuity and quantization in a single process, appears in the study of the conditional survival probabilities of a firm (the computation of the conditional survival probability of the firm from an investor's point of view, i.e., given the "investor information"). Callegaro and Sagna used a quantization procedure to analyze and compare the spread curves under complete and partial information in new and more general settings in their work on applications to credit risk of optimal quantization methods for nonlinear filtering. The theory of quantization probability they used was based on an earlier study of local quantization behavior of absolutely continuous probabilities [14]. This study analyzes the L^r quantization error estimates for $L^r(P)$ codebooks for absolutely continuous probabilities P and Voronoi partitions satisfying specific conditions. But the origins of the theory developed there can be traced back to electrical engineering and image processing and in particular in digitizing analog signals and compressing digital images [15]. Therefore, in the heart of the study of survival probabilities we find a theory for the quantization as analog-to-digital conversion and as data compression. Analog signal is a continuous signal which transmits information as a response to changes in physical phenomenon and uses continuous range of values to represent information, where digital signals are discrete time signals generated by digital modulation and use discrete or discontinuous values to represent information. The quality of a quantizer can be measured by the goodness of the resulting reproduction of a signal in comparison to the original. This is accomplished with the definition of a distortion measure that quantifies cost or distortion resulting from reproducing the signal, and the consideration of the average distortion as a measure of the quality of a system, with smaller average distortion meaning higher quality [15]. The design and analysis of practical quantization techniques can be tracked in three paths [15]:

- Fixed-rate scalar quantization, which adds linear processing to scalar quantization in order to exploit source redundancy, and variable-rate quantization (it uses Shannon's lossless source coding techniques [42] to reduce rate). Lossless codes were originally called noiseless.
- Vector quantization, including the seminal work of Shannon and Zador, in which vector quantization appears more to be a paradigm for analyzing the fundamental limits of quantizer performance than a practical coding technique.

11.2.4 Life of a Cell as Prey and Molecules as Predators

The predator–prey model is interesting in this context both in terms of modeling the different pathways of cell killing and in terms of time-domain study of treatment of the cells. The framework of a predator–prey interaction is employed using the concept of life unit cellular density as the prey and molecules inducing potential cell death as the predators. Hybridization by normalization of all concentrations underlies the predation. Each isolated pathway of cell death operates according to the underlying biochemical reaction or signaling mechanism. The resulting cell death is obtained by coupling of all death pathways through molecular effectors. Some molecules contribute to both cell proliferation and cell death. Cancer cells suppress cell apoptosis activities, but they do not disable the entire signaling cascade [33] which is evidence of the existence of molecules that contribute to both processes of cellular proliferation and cell death. Bcl2 family members are some of these molecules that have a dual role in cell death. In particular, the Bcl2 family members known as BH3-only molecules, lacking domains BH1, are known to have a pro-apoptotic character. Some members of this subfamily are Bad, Bid, $\mathrm{Bim}_{\mathrm{EL}}$, Bmf, and Mcl-$_{1\mathrm{S}}$. Some of the anti-apoptotic members of the Bcl-2 family are Bax, Bok/Mtd, Bak, and Bcl-$_{\mathrm{XS}}$ and they all contain at least one BH1 and BH2 domain. The pro-apoptotic Bcl2 members have been considered to play a role in the coupling of apoptosis and cell cycle arrest [33].

Holling was able to experimentally verify and derive the form of this nonlinear response, which is the famous "disk equation," identical to the Michaelis Menten term of enzyme kinetics,

$$\frac{Vx}{K+x} \tag{11.1}$$

where in the predator–prey context V is the maximum predator attack rate, K is prey density where the attack rate is half-saturated. To accommodate the difference in the time scale of predators (fast behavioral time scale) and the prey population (usually slower time scale), and to overcome the problem of incongruent time scaling, the nonlinear functional response should be expressed in terms of the ratio of prey to predator [4]. This expression, when inserted in the prey equation, solves the paradoxes of enrichment and biological control [3, 4].

The effect of potentially competing predators on a single limited prey has been studied extensively [13, 17, 35]. For example, it has been shown that if the interference coefficient between two predators is small, then the winner competes its rival successfully and if the interference coefficient is large enough, then the competition outcome depends on the initial population of predator species [19]. As mentioned earlier, a suitable predator–prey theory should be based on the so-called ratio-dependent theory, in which the per capita prey growth rate should be function of the ratio of prey to predator abundance, and should have the form of an empirical Michaelis-Menten ratio-dependent predator–prey functional response system. With this in mind we write the equations:

$$\frac{\mathrm{d}N}{\mathrm{d}t} = r\begin{bmatrix}\text{Proliferation}\\ \text{and}\\ \text{death}\\ \text{factors}\end{bmatrix} = \sum_{i=1}^{n} \frac{V_i(\text{inducer}_i)}{K_iN + w_i(\text{inducer})_i} \tag{11.2}$$

The double role of factors such as the p53 molecule and hypoxia points to internal system instability of the precise prediction of the deterministic cell death pathways as described by the biochemical and signaling differential equations. The p53 regulation in the modeling scheme we use is described in the differential equations (downregulation of Bcl2, etc.). The observed quiescence and cell cycle arrest, as well as the maintenance of the cell cycle by molecules of p53 during the attack of the death-inducing molecules (which might also be triggered by p53), we conjecture that it is a net effect of reproduction and loss of survival units which happens during the attack due to treatment, and is independent of the biochemical effects of the death inducers, but is rather a “fake death” (quiescence) mechanism of cancerous cells, that decrease their cell survival units as manifested by the arrest of the cell cycle, and at the same time preserve the wholeness of their viability and cell life, contained in its fullness in each remaining life unit.

The use of a predator–prey model (a continuous model used for the simulation of discrete population dynamics) for the modeling of survival probability (a continuous variable) suggests the quantization of survival probability. Indeed, the quantization of probability has been proposed by several other authors [1, 7, 18, 30–32, 36–38]. The existence of the “chance quantum” implies several axioms consequences according to Goldsmith [14], such as: (a) if an event has a calculated probability of less than one chance quantum it will not occur (b) for an event having an appreciable probability (equivalent to many chance quantums), a change in surrounding conditions leading to a computed change in probability of less than one chance quantum will in fact cause no change in the probability of the event. This indicates the existence of probability thresholds, which in the case of a cell could be interpreted as lower survival thresholds for the survival threshold, below which the damaged cell no longer has the potential to resist or to repair. For the upper thresholds, the cell will be able to recover. The idea of quantization of probability has been used in applications of linear circuits, in particular in the study of circuit analysis and methods needed to adequately predict circuit performance as a function of components tolerances (standard deviations from the mean). To obtain maximum component efficiency the quantized probability design (QPD) was developed in [7], a statistical method for predicting the tolerance limits for a circuit, where the parameters are weighted according to their effect on the circuit. In [23], the analog of the bathtub that is filling with discrete drops of water, which we perceive as a continuous flow, is considered. While the level changes while an observer is out of the bathroom, and the changed level appears as a discrete step, rather than a continuous curve, there is still a continuous flow of discrete drops of water while the observer is gone. The changed level and its stepped appearance do not make it a discrete event system [23]. Adopting this logic, we replace the bathtub by the biochemistry of a cell, the drops of water by discrete quantized units of cell survival probability, and the total water level in the bathtub by the continuous cell survival probability (biochemical life of the cell), which now is the sum of discrete quantized cell survival probabilities. This results in a duality in the perception of the nature of probability of survival of the cell. This allows the interpretation of survival probability as a discrete population, whose dynamics can be modeled using continuous equations; in particular predator–prey equations where the prey is the life of the cell and the predator is the cell death effectors.

This is a possible framework that may be adopted to study and analyze the process of cell survival during treatment. This suggests an organic connection among an axiomatic system foundation, a predator–prey rate equation, and information theoretic signal processing (Figs. 11.3 and 11.4).

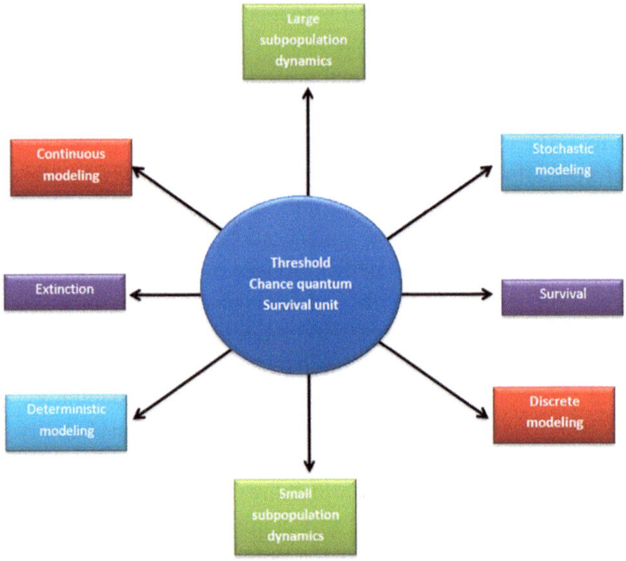

Fig. 11.3 The existence of threshold as the link between different aspects of the modeling process

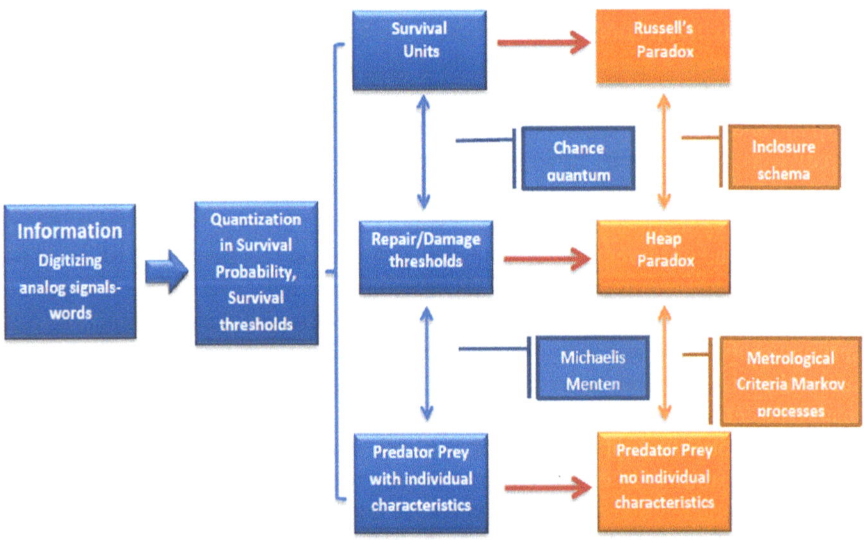

Fig. 11.4 Logical design of the steps that may lead to an organic connection of different components of modeling (information, axioms, rate equations) as the new approach to current issues in the foundations of mathematics

11.2.5 Quantized Protoplasm and Nanoprotoplasm Assembler

The suggested framework of an axiomatic interpretation of the survival of a cell has similarities with the theory of Quantized protoplasm. Each water molecule represented by the formula, H_2O, contains two hydrogen atoms and one oxygen atom, arranged in such a manner that one end of the water molecule is positively charged while the other end is negatively charged. In other words, each water molecule may be seen as a dipolar molecule [28]. In the framework of the new theory of Association Induction (AI) by G. Ling, induction (the mutual electrical polarization that results in induced dipole moments, either through interaction with a neighboring water molecule) is not present only between pairs of water molecules but among all the interacting protein-water-small molecules/ion assembly that makes up the living cell. This is what makes the different components of the cell *functionally coherent and discrete cooperative assemblies*. Cardinal adsorbents are defined as the biologically potent molecules that demonstrate an on or off action and shift an assembly from one resting-living-cooperative state to an active-living-cooperative state. Electrical polarization, or induction, thus brings about the association of all or virtually all water molecules in the cells [28]. If alternating positively charged P sites and negatively charged N sites are arranged in two dimensions at suitable distance apart like a checkerboard (an NP system), or if two such NP surfaces are face-to-face in close juxtaposition (an NP-NP system) or if alternating N and P sites are carried on linear chains among a matrix of similar chains (an NP-NPNP system), the interaction with and among the water molecules will be intensified. The interplay of induction and association can produce a stable and yet highly flexible, three-dimensional dynamic structure of polarized multi-layers of water molecules [26]. The theory of polarized multilayers as well as the association induction theory is a continuation of the studies of Thomas Huxley (1825–1895) that suggested that the living cell is the basic unit of all life, and that the substance of living cells called the "protoplasm" is "the physical basis of life." In these two complementary theories, being alive means that the components of the cell substance, proteins, water, and small molecules and ions are associated in a specific spatial relationship and in the high (negative) energy-low entropy state, called the living state (a cooperative state that allows for neighbor-to-neighbor electronic interaction among the individual elements). This idea leads to a view of the living cell as essentially an electronic machine, where the electronic perturbations are not carried out through long-range ohmic conduction of free electrons along electric wires but by a falling-domino-like propagated short-range interaction. In the dead state, water and ions are to a large extent liberated and exist as free water and free ions, with a large entropy gain. In death, the proteins enter an internally neutralized state [27]. The minimal structural unit of protoplasm that preserves the basic physical properties of the whole living cell is constituted by protein molecules with bound ATP, water, and potassium ions and the vital activity of the cell is reduced to transitions between two states[5]:

$$\text{protein}(\text{ATP})_m(\text{H}_2\text{O})_n(K^+)_q \leftrightarrow \text{protein} + mADP + mPi + n\text{H}_2\text{O} + qK^+ \quad (11.3)$$

We need to mention here that the goal of AI hypothesis is to interpret all microscopic cell physiological manifestations in terms of properties and activities of microscopic molecules, atoms, ions, and electrons. In this direction, cell nuclei, cell membranes, and other subcellular structures are made up of different varieties of macroscopic protoplasm. Although there is great diversity in form and function among these components, all macroscopic protoplasms have one thing in common: They all comprise a vast number of similar microscopic units called nanoprotoplasm (NP) which is claimed to be the smallest unit and ultimate physical basis of life [25]:

$$\text{nanoprotoplasm} = (A_1 B_m C_n \ldots Z_1)(\text{H}_2\text{O})_{a\times 1,000}(K^+)_{b\times 10}(\text{ATP})_c \quad (11.4)$$

where a, b, c, l, m, n are all positive integers and each of the symbols $A, B, C, \ldots, Z$ represents a different component protein. NP unit may contain just a single protein molecule characteristic of that particular protoplasm like (red blood cell cytoplasmic protoplasm), and in more complex protoplasm, each NP unit may contain two or more protein molecules of different kinds and number [25]. As an example, in the case of a red blood cell, the nanoprotoplasm unit is composed of one hemoglobin molecule, and contains some 7,000 water molecules, 20 K^+, and a single molecule of ATP, all directly or indirectly attached to the single hemoglobin molecule: $(Hb)_1(\text{H}_2\text{O})_{7,000}(K^+)_{20}(\text{ATP})_1$ A quick calculation shows that for a spherical nanoprotoplasm unit model, the diameter would be in the range of nanometers (roughly 8.6 nm) [25]. The fundamental characteristics of the dynamic structure (of chemical entities) shared by all nanoprotoplasm can be summarized as [25]:

- The universal possession of a long and a partially resonating polypeptide chain.
- The universal possession of two kinds of proximal functional groups: the β- and γ-carboxyl groups and the backbone NHCO groups and their access to alternative partners, K+, Na +, or fixed cations (for the β- and γ-carboxyl groups) and CONH groups belonging to the third amino acid residues up and down the polypeptide chain or massive number of water molecules (for the backbone NHCO groups).
- The universal presence of principle, auxiliary, and pseudocardinal sites and the right kind of cardinal adsorbent for each of them.

The distinction among different units is associated with the type and sequential order of amino acid residues of their protein components, with the location a unit occupies in the cell, with the way these components associate with one another in

[5] http://vladimirmatveev.ru/mainprinciples.html.

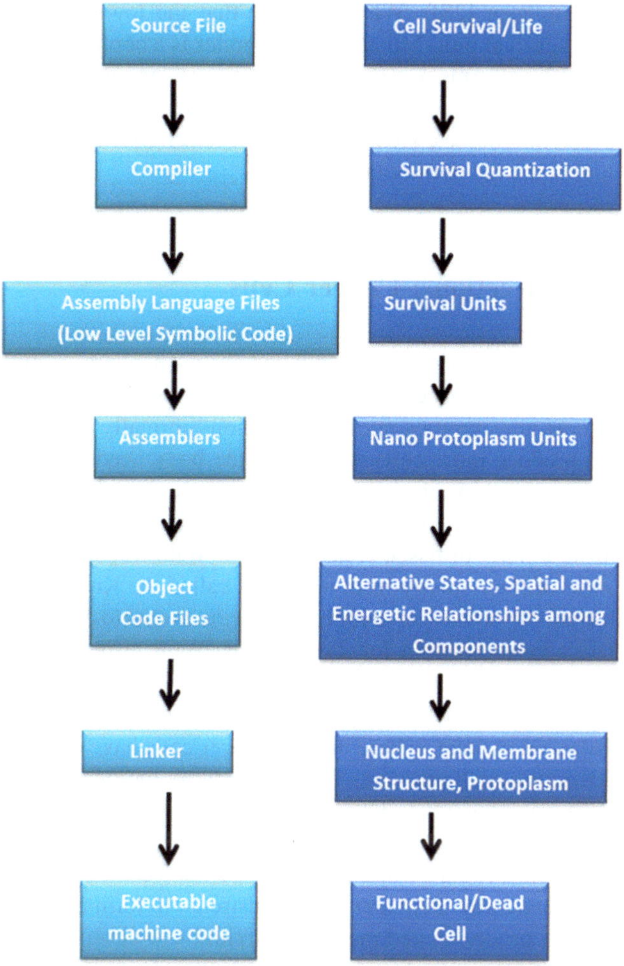

Fig. 11.5 Cell as a programmable device. The compiler translates the higher level language into assembly language. The assembler translates assembly language into object code. The linker builds an executable program from object modules and any library modules required

space and in energy [25]. It is considered that fixed β- and γ-carboxyl groups as *the highways of information and energy transfer*, where information in this context is any kind of event that affects the state of a dynamic system. Using terminology from computer science, a nanoprotoplasm assembly unit or better "nanoprotoplasm assembler" (Fig. 11.5) is a "utility" program that converts an assembly language (a low-level symbolic code, with low abstraction) into executable machine code (the biochemical type of nanoprotoplasm as given by the molecular formula above; what this code does is to alternate the chemical composition between two states, resting and active living state (or dead), and to determine the mutual spatial and energetic relationships among its components and in relation to the rest of the macroscopic protoplasm [25]). We conjecture that the low-level symbolic code corresponds to a survival unit. Moreover, a compiler is a computer program (or set of programs) that transforms a source code from one language to another. In the framework of a cell, this compiler represents the quantization of the cell survival to survival units, and the source code is cell survival of cell life. The physical basis for the survival unit is the nanoprotoplasm.

11.3 Future Directions: Cancer Cells

Cancer cells grow and divide at an unregulated pace. There are several differences between normal cells and cancer cells.[6] With respect to structure, normal cells have DNA in their genes and chromosomes that functions normally and they divide in an orderly way to produce more cells only when the body needs them. Cancer cells develop an aberrant DNA or gene structure or acquire abnormal numbers of chromosomes and continue to be created without control or order. This leads to excess cells from a mass of tissue called a tumor. With respect to energy, normal cells derive most of their energy from a process called the Krebs cycle and only a small amount from the process of glycolysis, and the means to derive these energies is oxygen. The opposite is true for cancer cells. With respect to blood vessels, normal cells have a built-in blood vessel system, something that cancer cells are lacking. With respect to functions, normal cells have enzymes and hormones that behave in a balanced manner, where instead cancer cells have either overactive or underactive enzymes or hormones. With respect to tumors, benign tumors of normal cells are not cancerous. They do not invade nearby tissues or spread to other parts of the body, can be removed, and are not a threat to life. Malignant tumors of cancer cells are cancerous and can invade and damage nearby tissues and organs and can break away and enter the bloodstream to form new tumors in other parts of the body, a process called metastasis. With these observations in mind, the modeling techniques we introduced may be (in the frame of cellular automata) suitable for testing of several hypotheses such as:

- Molecular species function as signaling molecules in many aspects of growth factor-mediated angiogenesis. Changes in such concentrations such as oxygen regulate neo-vascularization through induction of vascular endothelial growth factors (VEGF).
- Postoperative radiation therapies cause oxygen-related stimulation of immune response that under certain conditions can provoke tumor remission.
- PDT produces its tumoricidal effect through the generation of certain molecular species, which are toxic to cells and might also lead to destruction of the tumor microvasculature.

[6] Healthy Cells vs. Cancer Cells, A.P. John Institute for Cancer Research https://www.apjohncancerinstitute.org/frequently-asked-questions/healthy-cells-vc-cancer-cells.

This present modeling approach can be developed further through coupling with existing models of tumor neovascularization and the oxygen-regulated tumor-immune dynamics with angiogenesis taken into account, to study unquantified effects in the tumor macroenvironment with a small remnant of tumor tissue left after surgical resection as the initial condition. Moreover, the problem can be enriched by other parameters such as the effect of heating tissue, which decreases the viscosity of fluid elements, increases metabolic rate, increases blood flow which assists in the reduction of swelling, and stimulates the immune system. All these factors might play a significant role in the final outcome.

Conclusion

The idea of quantization of survival probability is reflected in a predator–prey form of the cell survival equation. Starting from cell survival differential equation, we can identify the similarities to the dynamic energy budget models that study the "strategy" that an organism might develop to optimize its overall fitness, measured, for example, by a net reproductive output. The probability of survival for an individual organism is determined by the principal hazard for most creatures which is predation, and the risk of predation is dependent on the size of the organism. This is the link between predator–prey theory and the idea of the development of a strategy.

A new set theoretic approach is also introduced through the definition of cell survival units or cell survival units indicating the use of "proper classes" according to the Zermelo–Fraenkel set theory and the axiom of choice, as the mathematics appropriate for the development of biological theory of cell survival.

Our ultimate goal is to initiate a program that will enable a physician to evaluate photochemical tumor treatment and to better design a patient-specific therapy to achieve maximum destruction of the tumor and injury minimization of healthy tissue by controlling time and drag concentrations in tissue. The hard work is yet to be done: researchers will need to formulate and verify models, estimate kinetic parameters, make non-obvious predictions, and test them by quantitative experimental measurements [46]. It is a matter of time before effective, integrated models of regulatory networks in cancer cells are used as an informational supplement for the next wave of experiments and therapies.

Mathematics strongly prizes rigor and precision. Mathematical fact is immutable, and successful mathematical theories have lifetimes of hundreds or thousands of years. By contrast, most of our knowledge of biological systems is recent, and most biological theories evolve rapidly. Nonetheless, the interface between mathematics and biology has initiated and fostered new mathematical areas. This report highlights areas of mathematics that require future development and extensions in order to be useful for some areas of biology modeling. Meaningful descriptions of fundamental biological principles require the development of new mathematical tools.

References

1. Aaron MR, Simon MK (1966) Approximation of error probability in a regenerative repeater with quantized feedback. Bell Syst Tech J 45:1845
2. Anand P, Kunnumakara AB, Sundaram C, Harikumar KB, Tharakan ST, Lai OS, Sung BY, Aggarwal BB (2008) Cancer is a preventable disease that requires major lifestyle changes. Pharm Res 25:2097–2116
3. Arditi R, Berryman AA (1991) The biological-control paradox. Trends Ecol Evol 6:32–32
4. Arditi R, Ginzburg LR (1989) Coupling in predator prey dynamics—ratio-dependence. J Theor Biol 139:311–326
5. Balazsi G, Van Oudenaarden A, Collins JJ (2011) Cellular decision making and biological noise: from microbes to mammals. Cell 144:910–925
6. Bartlett MS (1957) On theoretical models for competitive and predatory biological systems. Biometrika 44:27–42
7. Burns RC, Lawson AD (1964) Quantized probability circuit design principles applied to linear circuits. IEEE Trans Reliab R 13:16
8. Cantor G (1932) Gesammelte Abhandlungen mathematischen und philosophischen inhalts (Ernst Zermelo ed.). Springer, Berlin, Includes excerpts of his correspondence with Dedekind (1899)
9. Champagnat N (2006) A microscopic interpretation for adaptive dynamics trait substitution sequence models. Stochastic Processes Appl 116:1127–1160
10. Chen WW, Niepel M, Sorger PK (2010) Classic and contemporary approaches to modeling biochemical reactions. Genes Dev 24:1861–1875
11. Chesson P (1978) Predator–prey theory and variability. Annu Rev Ecol Systemat 9:323–347
12. Colyvan M (2009) Vagueness and truth. In: Dyke H (ed) From truth to reality: new essays in logic and metaphysics. Routledge, New York, NY
13. Ferriere R, Bronstein JL, Rinaldi S, Law R, Gauduchon M (2002) Cheating and the evolutionary stability of mutualisms. Proc Biol Sci 269:773–780
14. Freedman HI, Waltman P (1984) Persistence in models of 3 interacting predator–prey populations. Math Biosci 68:213–231
15. Goldsmith A (1943) Quantized probability. Phys Rev 64:376
16. Graf S, Luschgy H, Pages G (2012) The local quantization behavior of absolutely continuous probabilities. Ann Probab 40:1795–1828
17. Gray RM, Neuhoff DL (1998) Quantization. IEEE Trans Inform Theor 44:2325–2383
18. Hâajek P, Pudlâak P (1993) Metamathematics of first-order arithmetic. Springer, Berlin
19. Harrison GW (1979) Global stability of predator–prey interactions. J Math Biol 8:159–171
20. Hoge H (1976) Calculation of error probability in regenerative systems with quantized feedback. Aeu Int J Electron Comm 30:130–133
21. Hsu SB (1981) On a resource based ecological competition model with interference. J Math Biol 12:45–52
22. Kalyanasundaram S, Chong EKP, Shroff NB (2004) Markov decision processes with uncertain transition rates: sensitivity and max hyphen min control. Asian J Contr 6:253–269
23. Kleinbaum DG (2005) Survival analysis; a self-learning text. Springer, New York, NY
24. Kossak R, Schmerl JH (2006) The structure of models of Peano arithmetic. *Oxford logic guides 50*. Clarendon, Oxford
25. Kuipers B (1989) Qualitative reasoning—modeling and simulation with incomplete knowledge. Automatica 25:571–585
26. Leslie PH (1958) A stochastic model for studying the properties of certain biological systems by numerical methods. Biometrika 45:16–31
27. Ling G (2007) Nano-protoplasm: the ultimate unit of life. Physiol Chem Phys Med NMR 39:111–234
28. Ling GN (1965) Physical state of water in living cell and model systems. Ann N Y Acad Sci 125:401
29. Ling GN (1988) A physical theory of the living state—application to water and solute distribution. Scanning Microsc 2:899–913

30. Ling GN (1994) The new cell physiology—an outline, presented against its full historical background, beginning from the beginning. Physiol Chem Phys Med NMR 26:121–203
31. link G, Ebrary INC (2004) One hundred years of Russell's paradox mathematics, logic, philosophy. *De Gruyter series in logic and its applications 6*. Walter de Gruyter, New York
32. Lomsadze YM (1962) Relativistically invariant formulation of theory of quantized probability amplitude field. Nucl Phys 37:147
33. Lomsadze YM, Krivsky IY, Khimitch IV (1962) A consistently relativistically invariant formulation of the quantised probability-amplitude field theory. Phys Lett 2:80–81
34. Lomsadze YM, Krivsky IY, Khimitch IV (1962) A possible experimental determination of the universal constant-I in the quantised probability amplitude field theory. Phys Lett 2:218–219
35. Maddika S, Ande SR, Panigrahi S, Paranjothy T, Weglarczyk K, Zuse A, Eshraghi M, Manda KD, Wiechec E, LOS M (2007) Cell survival, cell death and cell cycle pathways are interconnected: implications for cancer therapy. Drug Resist Updat 10:13–29
36. Masutani K (1993) Effects of survival thresholds upon one-dimensional dynamics of single-species populations. Bull Math Biol 55:1–13
37. Mitra D, Mukherjee D, Roy AB, Ray S (1992) Permanent coexistence in a resource-based competition system. Ecol Model 60:77–85
38. Ohta M, Hatakeyama K, Nishimura M (1979) Unified study of the multivariate joint probability function of the state variables with quantized levels for a stochastic environmental system with discrete data—theory and experiment. J Sound Vib 66:75–89
39. Ohta M, Yamaguchi S, Katho Y (1986) Multivariate probability expression of quantized state variables for a nonstationary system and its application to an acoustic environment. J Sound Vib 104:465–482
40. Ohta M, Yamaguchi S, Mitani Y (1984) A precise noise probability estimation from roughly observed data with quantized levels. J Acoust Soc Am 76:122–127
41. Perthame B, Gauduchon M (2010) Survival thresholds and mortality rates in adaptive dynamics: conciliating deterministic and stochastic simulations. Math Med Biol 27:195–210
42. Priest G (2002) Beyond the limits of thought. Clarendon/Oxford University Press, Oxford
43. Quine WV (1951) On the consistency of new foundations. Proc Natl Acad Sci U S A 37:538–540
44. Schröder E (1890–1905) Vorlesungen über die algebra der logik, 3 vols. B.G. Teubner, Leipzig, Reprints: (1966) Chelsea; (2000) Thoemmes Press
45. Shannon CE (1948) A mathematical theory of communication. Bell Syst Tech J 27:623–656
46. Simmons K (2000) Sets, classes and extensions: a singularity approach to Russell's paradox. Phil Stud 100:109–149
47. Specker EP (1953) The Axiom of Choice in Quine New Foundations for Mathematical Logic. Proc Natl Acad Sci U S A 39:972–975
48. Takahashi K, Yamanaka S (2006) Induction of pluripotent stem cells from mouse embryonic and adult fibroblast cultures by defined factors. Cell 126:663–676
49. TYSON JJ, BAUMANN WT, CHEN C (2011) Dynamic modelling of oestrogen signalling and cell fate in breast cancer cells. Nat Rev Cancer 11(7):523–532
50. van Heijenoort J (1967) From Frege to Gödel; a source book in mathematical logic, 1879–1931. Harvard University Press, Cambridge
51. von Neumann J (1997) An axiomatization of set theory. Aut Aut 280:107–123
52. Vopenka P (1979) Mathematics in the alternative set theory. B. G. Teubner, Leipzig
53. Wheatley N (2011) A sorites paradox in the conventional definition of amount of substance. Metrologia 48:L17–L21
54. Yu RC, Rappaport SM (1997) A lung retention model based on Michaelis-Menten-like kinetics. Environ Health Perspect 105:496–503
55. Zehna PW, Johnson RL (1962) Elements of set theory. College mathematics series. Allyn and Bacon, Boston, MA
56. Zhou HX, Rivas GN, Minton AP (2008) Macromolecular crowding and confinement: biochemical, biophysical, and potential physiological consequences. Annu Rev Biophys 37:375–397

Chapter 12
Possible Role of Resveratrol Targeting Estradiol and Neprilysin Pathways in Lipopolysaccharide Model of Alzheimer Disease

Nesrine S. El-Sayed and Yasmeen Bayan

Abstract Alzheimer's disease (AD) is an irreversible, progressive neurodegenerative brain disease that slowly destroys memory and thinking skills. It is the most common cause of dementia among older people. One of the most important hallmarks of AD is the presence of amyloid beta (Aβ) peptide in the brain that suggests that it is the primary trigger for neuronal loss. Herbal extracts have been studied over the years for their potential therapeutic effect in AD. Resveratrol (RSV), one of the most important phytoestrogens, is considered to be useful as estrogen plays an important role in AD. One of the most important amyloid degrading enzymes is neprilysin (NEP), which plays a major role in degrading Aβ, and mainly affected by estrogen. So, the aim of the present study is investigating the possible role of resveratrol in lipopolysaccharide model of AD and the implication of its possible role in regulating the estradiol and neprilysin pathways. Mice were divided into four groups: Control group (0.9 % saline), LPS group (0.8 mg/kg i.p once), Treatment group with RSV (mice were once injected with LPS then after 30 min given a dose of {4 mg/kg} RSV for 7 days), and RSV group only (mice received 4 mg/kg i.p for 7 days only). After 7 days mice were subjected to different behavioral tests using Y-maze, object recognition test, and open field tests. Estradiol and NEP level were measured using ELISA kit. Results showed RSV was able to reverse the decline in different types of memory (working, nonspatial, and locomotor functions) caused by LPS induction in mice. Moreover RSV was able to significantly increase both the estradiol level and NEP level and that may have a great role to decrease Aβ deposition as it has been confirmed that there is a link between NEP and estradiol level; by upregulation of estradiol level this consequently leads to increase in the

N.S. El-Sayed (✉)
Pharmacology and Toxicology Department, Faculty of Pharmacy and Biotechnology, German University in Cairo, Cairo, Egypt
e-mail: nesrine.elsayed@guc.edu.eg

Y. Bayan
German University in Cairo, Cairo, Egypt
e-mail: yuan_11_11@hotmail.com

P. Vlamos, A. Alexiou (eds.), *GeNeDis 2014*, Advances in Experimental Medicine and Biology 822, DOI 10.1007/978-3-319-08927-0_12

level of NEP level, and by increasing the NEP level in brain, this lead to decrease in Aβ deposition and enhancing its degradation by NEP

Keywords Alzheimer disease • Resveratrol • Neprilysin • Lipopolysaccharide • Estrogen

Abbreviations

(Aβ)	Amyloid beta peptides
AD	Alzheimer's disease
DES	Diethylstilbestrol
DR	Discrimination ratio
i.p.	Intraperitoneal
LPS	Lipopolysaccharide
MAP	Mean alternation percentage
NEP	Neprilysin
RSV	Resveratrol

12.1 Introduction

Alzheimer's disease (AD) is a progressive neurodegenerative disease that is characterized by impairment of memory and disturbance in at least one of the thinking functions. It is considered the most common cause of dementia, affecting more than 25 million people worldwide [5] and considered the fifth leading cause of death for those aged more than 65 years. The most common hypothesis behind AD is the deposition of neurotic plaques extracellularly and neurofibrillary tangles intracellularly in the brain of AD patients. Several major theories are involved in the pathogenesis of AD, such as neurotic plaques hypothesis [45], tauopathy [30], inflammation [37], and oxidative stress [41].

Neurotic plaque hypothesis is considered the primary hallmark of AD in which APP (protein normally found in brain essential for the growth and survival of the neurons) is cleaved by mutant form of β-secretase then followed by γ- secretase which results in generation of the Aβ peptide which aggregates and deposit on the surface of the neurons [32].

Different mechanisms were implemented in order to decrease the level of beta amyloid plaques, one of them is altering catabolism of Aβ levels in the brains of AD, by using many proteases and peptidases that have the capability of cleaving Aβ either in vitro or in vivo. The following proteinases have the abilities of degrading Aβ peptide. One of the primary degrading enzyme which targets the degradation of beta amyloid peptides is Neprilysin enzyme (NEP) which is found to be abundant in the brain areas vulnerable to amyloid plaque deposition [23]. It is considered the major amyloid degrading enzyme and plays a major role in the clearance of amyloid beta peptides (Aβ) from the brain [22, 36]. NEP was found to degrade both monomeric and oligomeric forms of Aβ40 and Aβ42 in intracellular and extracellular compartments of the brain [16, 56].

Estrogen deprivation has been implicated in the pathogenesis of AD [54]. Furthermore, there is a strong evidence that AD pathology and AD-related cognitive decline are strongly related to decrease in estrogen level [43, 53]. Estrogen plays an important role in protection against neuron cell death [48, 52] and has the ability of inhibition of different aspects of AD neuropathology including Aβ accumulation [4, 42].

Interestingly NEP was found to be mainly upregulated by estrogen hormone [19]. Moreover it has been proven that estrogen deprivation caused a lowering of NEP level and that estrogen replacement therapy restored NEP level to control values in ovariectomized rat brains. It has been speculated that estrogen regulates the expression of NEP [13]. Furthermore NEP activity was found to be lower in the hippocampus, cerebellum, and caudate of ovariectomized rats than in non-ovariectomized rats, and this effect can be reversed by exogenous 17β-estradiol [59].

Lipopolysaccharide (LPS) is a gram negative bacterial endotoxin that was found to induce neuroinflammation which exhibits AD-like features when systemically injected to rodents [7]. Moreover it has been proven that intraperitoneal (i.p.) injection of LPS induces cognitive impairment in mice as it induces memory loss and amyloidogenesis in vivo and in vitro as a consequence of systemic inflammation [46, 51] by production of proinflammatory cytokines such as IL-1β, IL-6, and TNF-α [38, 17].

Resveratrol (RSV) (3,4′,5-trihydroxystilbene), a secondary metabolite, belongs to a class of polyphenolic compounds called stilbenes [49]. It was first isolated from the roots of *Veratrum grandiflorum* [29]. Several plant species are known to produce RSV (classified as polyphenol) such as vine plant, peanuts, berries but was found to be in large amounts in skin of red grapes [11, 40].

RSV possesses many biological activities that are protective against atherosclerosis, including antioxidant, anti-inflammatory, antiplatelet, and vasorelaxant activities, and has been used in applying cardioprotective effect [35]. In addition, it is identified as a cancer chemotherapeutic agent [26] and has recently been reported to mimic effects of dietary restriction and extends the life span of lower organisms [18, 58].

It has been demonstrated that RSV can lessen amyloid-β peptide-induced toxicity, protect against cerebral ischemic injury, and shield neurons from induced excitotoxicity [28]. Moreover, RSV promotes the non-amyloidogenic pathway and enhances the cleavage of the amyloid precursor protein. Furthermore it enhances clearance of amyloid beta peptides, and reduces neuronal damage [31].

One of the important features that RSV has been characterized with is being a phytoestrogen on the basis of its ability to bind to estrogen receptors α and β (ERs α and ER β) and elicits similar responses to endogenous estrogens. Moreover, the chemical structure of RSV is very similar to that of the synthetic estrogen agonist, diethylstilbestrol (DES). RSV is considered an estrogen agonist [6]. It was found that RSV in particular acts on estrogen receptor beta which has a major function in the cognitive processes, so by binding of RSV to estrogen receptor beta, this results in improvement of cognitive impairment which results from AD [10].

So due to the structural similarity between RSV and estradiol this study was conducted to investigate the strong correlation between RSV and NEP in the degradation of amyloid beta.

12.2 Methodology

After 1 week of acclimatization, mice were randomly allocated to four groups each consisting of 10–12 mice. One group served as normal and received (saline 0.9 % i.p.) once.

The second group mice were injected with LPS (0.8 mg/kg i.p in saline) once to induce systemic and neuroinflammation and then were subjected to the tests after 7 days [1, 47, 14]. The third group mice were injected with LPS (0.8 mg/kg i.p in saline) once, and after 30 min mice were injected with RSV (4 mg/kg in saline) for 7 days [8, 9]. RSV was dissolved in the least amount of DMSO and completed to the final volume with saline [12, 24]. Animals were subjected to the tests after the seventh day of all injections. The fourth group mice were injected with only RSV (4 mg/kg i.p in saline) and were subjected to the tests after the seventh day of injections. All Animals were subjected to different behavioral tests such as open field, object recognition, and Y-maze behavioral tests at the end of the previously mentioned treatments, then they were sacrificed, and their brains were harvested. The brains were divided into two hemispheres and were frozen at 80 °C to be later homogenized and used for ELISA Kit.

12.3 Relation Between Resveratrol as a Phytoestrogen and Neprilysin in Lipopolysaccharide Induced Alzheimer's Disease

Behavioral tests were performed as a form of noninvasive method to detect changes in cognitive function after treating mice with LPS. Three behavioral tests were conducted: Y-maze test to detect working memory, open field test to measure locomotor functions, and object recognition test to measure nonspatial memory.

LPS-induced cognitive impairment was associated with a decrease in ambulation, rearing, and grooming frequencies compared to the normal animals in open field testing. It also resulted in significant decrease in working memory evaluated by significant decrease in MAP measured by Y-maze test. These results are in agreement with those of [14], in which administration of LPS once then testing

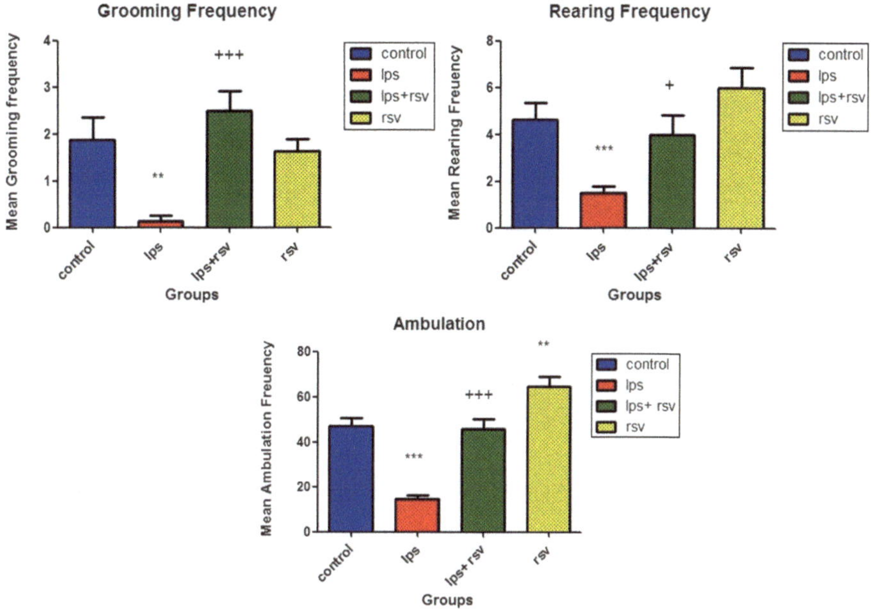

Fig. 12.1 Effects of LPS and resveratrol on the ambulation, grooming, and rearing frequencies

mice after 7 days from injection significantly reduced locomotor activity as well as reduced MAP (Figs. 12.1 and 12.2).

LPS-induced AD was associated with a decrease in nonspatial memory evaluated using new object recognition test. The results showed a significant decrease in the discrimination ratio compared to normal mice. This result is in agreement with those of [25] in which LPS injection showed significant nonspatial memory deficit evaluated by the impairment of object memory in new object recognition test. This object recognition memory deficit was evaluated to be due to systemic response induced by the i.p. injection of LPS (Fig. 12.3).

RSV, a polyphenolic compound found mainly in skin of red grapes, was found to induce in vivo protective properties against multiple illnesses, including cancer, cardiovascular disease, and ischemia, and was also found to confer resistance to stress and to extend life span [3].

Administration of RSV resulted in significant increase in locomotor activity evaluated by open field test. Our results are in agreement with a study in which RSV has showed that it can contribute to the preservation of cognitive function during aging as it was shown in the results that RSV was able to improve the locomotor function using open field test [44] (Fig. 12.1).

Administration of RSV resulted in significant increase in mean alternation percentage evaluated by Y-maze test. This result is in accordance with a study in which RSV was examined to see its effect on learning and working memory in Y-maze in diabetic rat; results showed that RSV elevated MAP significantly and improved learning and working memory [39] (Fig. 12.2).

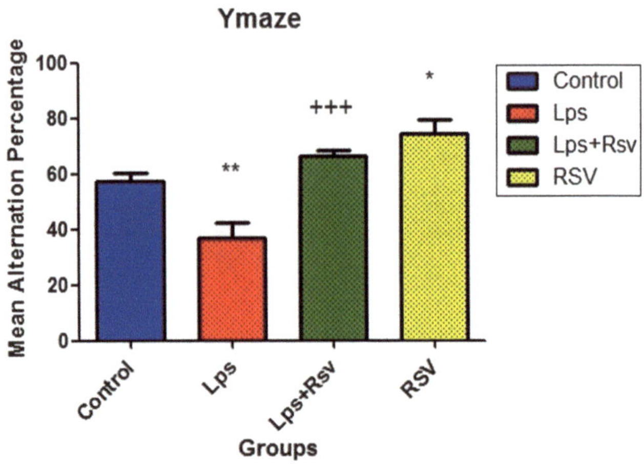

Fig. 12.2 Effect of LPS and resveratrol on the mean alternation percentage

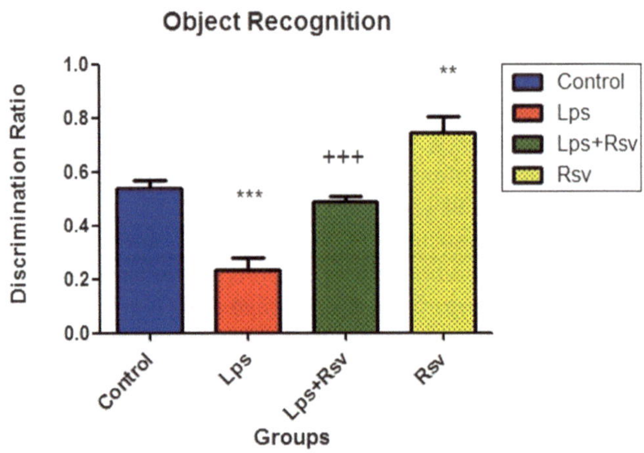

Fig. 12.3 Effect of LPS and resveratrol on the discrimination ratio

Administration of RSV resulted in significant increase in nonspatial memory evaluated by object recognition test. This result is in harmony with a study in which RSV has been administrated to assess its effect on subarachnoid and intracranial hemorrhage, ischemia, and stroke. Results showed that administration of RSV was able to improve nonspatial memory dramatically after trauma and that posttraumatic memory was restored to 66 % compared to uninjured control [50] (Fig. 12.3).

Previous studies proved that estrogen treatment inhibits microglial activation following exposure to the inflammatory stimuli (LPS) in LPS-treated mice, and that resulted in increasing cell viability and attenuating the inflammatory effect of LPS [55].

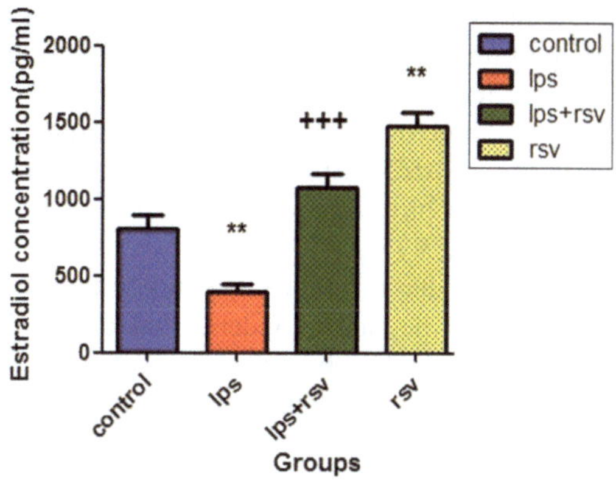

Fig. 12.4 Effect of LPS and resveratrol on estradiol level

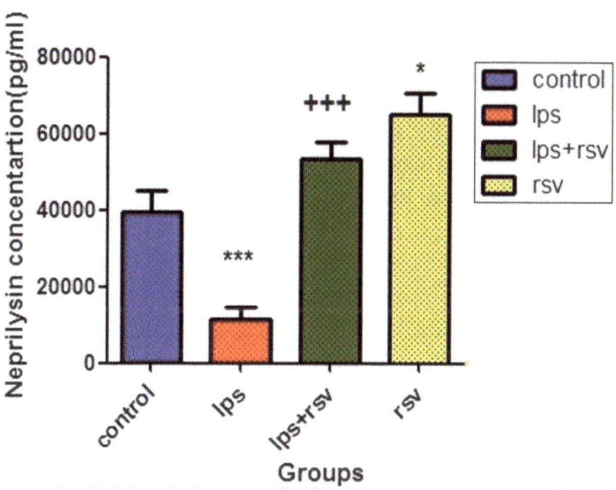

Fig. 12.5 Effect of LPS and resveratrol on neprilysin level

Administration of LPS (0.8 mg/kg i.p.) once significantly decreased estrogen level when measured by estradiol ELISA kit. These results are in agreement with a study done by [57] in which LPS and TNF-α suppress ovarian cell function and decreased estrogen level significantly (Fig. 12.4).

Administration of LPS (0.8 mg/kg i.p.) once significantly decreased NEP level when measured by NEP ELISA kit. These results are in agreement with a study done by [15] in which administration of LPS resulted in significant decrease in the level of NEP (Fig. 12.5).

RSV, a phytoestrogen, has been proven to be able to confer a significant improvement in spatial memory, and protect animals from Aβ-induced neurotoxicity. These neurological protection effects of resveratrol were associated with elevated estrogen level together with reduced iNOS level and oxidative stress [20].

Administration of RSV a phytoestrogen (4 mg/kg i.p) for 7 days either in the group of mice that were injected with LPS to induce AD or in the other group that was injected with RSV, both groups showed a very high significant increase in estrogen level when measured by estradiol ELISA kit based on its structural similarity to diethylstilbestrol. Similar to our results, different studies were conducted to prove that administration of RSV led to elevation in the level of estradiol [2, 27] (Fig. 12.4).

NEP is a zinc metalloendopeptidase that regulates the activity of a number of physiological peptides. Evidence has accumulated that NEP is involved in the clearance of amyloid beta peptides in the brain. Several studies provide evidence that show a strong relation between estrogen and NEP enzyme which is considered the major degrading enzyme for beta amyloid plaques in AD brain. Previous studies have shown that NEP gene responds to progesterone, androgen, and glucocorticoids, but primarily its activity is regulated by estrogen [21].

It has been proven that experimental depletion of endogenous NEP activity by pharmacological inhibitors or reduction of NEP expression by molecular approaches results in neural accumulation of Aβ [33].

Administration of RSV, a phytoestrogen, resulted in significant increase in NEP level when measured by NEP ELISA kit. These results are in agreement with a study by Marambaud et al. where RSV promoted the clearance of Alzheimer's disease amyloid beta peptides through upregulation of NEP level [34] (Fig. 12.5).

12.4 Results

Normal animals group were injected once with saline 7 days before subjecting them to open field testing (Group 1). Another group (4) was injected only with RSV (4 mg/kg) for 7 days and was subjected to open field after 7 days. Alzheimer's disease was induced in groups 2 and 3 by intraperitoneal injection of LPS (0.8 mg/kg) 7 days before subjecting them to open field testing. Group 3 was injected with RSV (4 mg/kg i.p.) for 7 days after 30 min from injection the LPS, and was subjected to open field test on the seventh day. Analysis was carried out using unpaired *t*-test to compare each two groups.

Each value represents Mean ± Standard Error of Mean. *Significantly different from Group 1 at $P < 0.05$. $^{+}$Significantly different from Group 2 at $P < 0.05$.

Normal animals group were injected once with saline 7 days before subjecting them to Y-maze testing (Group 1). Another group (4) was injected only with RSV (4 mg/kg) for 7 days and was subjected to Y-maze test after 7 days. Alzheimer's disease was induced in groups 2 and 3 by intraperitoneal injection of LPS (0.8 mg/kg) 7 days before subjecting them to Y-maze test. Group 3 was injected with RSV (4 mg/kg) for 7 days after 30 min from injection the LPS, and was subjected to Y-maze test on the seventh day. Analysis was carried out using unpaired *t*-test to compare each two groups.

Each value represents Mean ± Standard Error of Mean. *Significantly different from Group 1 at $P < 0.05$. $^{+}$Significantly different from Group 2 at $P < 0.05$.

MAP = Mean Alternation Percentage.

Normal animals group were injected once with saline 7 days before subjecting them to object recognition testing (Group 1). Another group (4) was injected only with RSV (4 mg/kg) for 7 days and was subjected to object recognition test after 7 days. Alzheimer's disease was induced in groups 2 and 3 by intraperitoneal injection of LPS (0.8 mg/kg) 7 days before subjecting them to object recognition test. Group 3 was injected with RSV (4 mg/kg) for 7 days after 30 min from injection the LPS, and was subjected to object recognition test on the seventh day. Analysis was carried out using unpaired *t*-test to compare each two groups.

Each value represents Mean ± Standard Error of Mean. *Significantly different from Group 1 at $P < 0.05$. +Significantly different from Group 2 at $P < 0.05$.

Discrimination Ratio = DR

Animals were divided into four groups; group1 is the normal animal group injected only with 0.9 % saline i.p once. Alzheimer's disease was induced in groups 2 and 3 by intraperitoneal injection of LPS (0.8 mg/kg i.p) once. Group 2 is the untreated group while the treated group is (group 3) that was injected with RSV (4 mg/kg i.p) for 7 days after 30 min from injection with LPS. Another group of mice injected with RSV only (4 mg/kg i.p) for 7 days (group 4). After performing the behavioral tests, brains were obtained and homogenized as 10 % homogenate in PBS, centrifuged, and supernatants were obtained. Fifty microliters of the supernatants from six samples of each group was diluted to 50 μl with assay buffer, added on it reaction mix and assayed. Then substituted by the absorbance in the straight line equation to get the concentrations and calculated as Sa/SV. Statistical analysis of estradiol concentrations was carried out using unpaired *t*-test to compare each two groups.

Each value represents Mean ± Standard Error of Mean. *Significantly different from Group 1 at $P < 0.05$. +Significantly different from Group 2 at $P < 0.05$.

Animals were divided into four groups; group1 is the normal animal group injected only with 0.9 % saline i.p. once. Alzheimer's disease was induced in groups 2 and 3 by intraperitoneal injection of LPS (0.8 mg/kg i.p) once. Group 2 is the untreated group while the treated group is (group 3) that was injected with RSV (4 mg/kg i.p) for 7 days after 30 min from injection with LPS. Another group of mice injected with RSV only (4 mg/kg i.p) for 7 days (group 4). After performing the behavioral tests, brains were obtained and homogenized as 10 % homogenate in PBS, centrifuged, and supernatants were obtained. Fifty microliters of the supernatants from six samples of each group was diluted to 50 μl with assay buffer, added on it reaction mix and assayed. Then substituted by the absorbance in the straight line equation to get the concentrations and calculated as Sa/SV. Statistical analysis of estradiol concentrations was carried out using unpaired *t*-test to compare each two groups.

Each value represents Mean ± Standard Error of Mean. *Significantly different from Group 1 at $P < 0.05$. +Significantly different from Group 2 at $P < 0.05$.

Conclusion

- LPS can be a perfect model for induction of AD in rodents.
- RSV can be a promising drug in management of AD.

References

1. Arai K et al (2001) Deterioration of spatial learning performances in lipopolysaccharide-treated mice. Jpn J Pharmacol 87(3):195–201
2. Basly JP et al (2000) Estrogenic/antiestrogenic and scavenging properties of (E)- and (Z)-resveratrol. Life Sci 66(9):769–777
3. Baur JA, Sinclair DA (2006) Therapeutic potential of resveratrol: the in vivo evidence. Nat Rev Drug Discov 5(6):493–506
4. Brinton RD (2001) Cellular and molecular mechanisms of estrogen regulation of memory function and neuroprotection against Alzheimer's disease: recent insights and remaining challenges. Learn Mem 8(3):121–133
5. Brookmeyer R et al (2011) National estimates of the prevalence of Alzheimer's disease in the United States. Alzheimers Dement 7(1):61–73
6. Bowers JL et al (2000) Resveratrol acts as a mixed agonist/antagonist for estrogen receptors alpha and beta. Endocrinology 141(10):3657–3667
7. Buxbaum JD et al (1992) Cholinergic agonists and interleukin 1 regulate processing and secretion of the Alzheimer beta/A4 amyloid protein precursor. Proc Natl Acad Sci U S A 89 (21):10075–10078
8. Canistro D et al (2009) Alteration of xenobiotic metabolizing enzymes by resveratrol in liver and lung of CD1 mice. Food Chem Toxicol 47(2):454–461
9. Chan V et al (2011) Resveratrol improves cardiovascular function in DOCA-salt hypertensive rats. Curr Pharm Biotechnol 12(3):429–436
10. Collins MA et al (2009) Alcohol in moderation, cardioprotection, and neuroprotection: epidemiological considerations and mechanistic studies. Alcohol Clin Exp Res 33(2):206–219
11. Das DK, Maulik N (2006) Resveratrol in cardioprotection: a therapeutic promise of alternative medicine. Mol Interv 6(1):36–47
12. Doubek J et al. (2005) Effect of stilbene resveratrol on haematological indices of rats. Acta Vet Brno roč. 74, č. 2, s. 205–208. ISSN 0001-7213
13. Dubrovskaia NM et al (2009) Changes in the activity of amyloid-degrading metallopeptidases leads to disruption of memory in rats. Zh Vyssh Nerv Deiat Im I P Pavlova 59(5):630–638
14. El Sayed NS, Kassem LA, Heikal OA (2009) Promising therapy for Alzheimer's disease targeting angiotensinconverting enzyme and the cyclooxygense-2 isoform. Drug Discov Ther 3(6):307–315
15. Hashimoto S et al (2010) Expression of neutral endopeptidase activity during clinical and experimental acute lung injury. Respir Res 11:164
16. Hauss-Wegrzyniak B, Wenk GL (2002) Beta-amyloid deposition in the brains of rats chronically infused with thiorphan or lipopolysaccharide: the role of ascorbic acid in the vehicle. Neurosci Lett 322(2):75–78
17. Heo S-K et al (2008) LPS induces inflammatory responses in human aortic vascular smooth muscle cells via Toll-like receptor 4 expression and nitric oxide production. Immunol Lett 120 (1–2):57–64
18. Howitz KT et al (2003) Small molecule activators of sirtuins extend Saccharomyces cerevisiae lifespan. Nature 425(6954):191–196

19. Huang J et al (2005) Binding of estrogen receptor beta to estrogen response element in situ is independent of estradiol and impaired by its amino terminus. Mol Endocrinol 19 (11):2696–2712
20. Huang T-C et al (2011) Resveratrol protects rats from Aβ-induced neurotoxicity by the reduction of iNOS expression and lipid peroxidation. PLoS One 6(12):e29102
21. Huang J et al (2004) Estrogen regulates neprilysin activity in rat brain. Neurosci Lett 367 (1):85–87
22. Iwata N et al (2000) Identification of the major Abeta1-42-degrading catabolic pathway in brain parenchyma: suppression leads to biochemical and pathological deposition. Nat Med 6 (2):143–150
23. Iwata N et al (2002) Region-specific reduction of a beta-degrading endopeptidase, neprilysin, in mouse hippocampus upon aging. J Neurosci Res 70(3):493–500
24. Iwuchukwu OF, Nagar S (2008) Resveratrol (trans-resveratrol, 3,5,4′-trihydroxy-trans-stilbene) glucuronidation exhibits atypical enzyme kinetics in various protein sources. Drug Metab Dispos 36(2):322–330
25. Jacewicz M et al (2009) Systemic administration of lipopolysaccharide impairs glutathione redox state and object recognition in male mice. The effect of PARP-1 inhibitor. Folia Neuropathol 47(4):321–328
26. Jang M et al (1997) Cancer chemopreventive activity of resveratrol, a natural product derived from grapes. Science 275(5297):218–220
27. Juan ME et al (2005) trans-Resveratrol, a natural antioxidant from grapes, increases sperm output in healthy rats. J Nutr 135(4):757–760
28. Karuppagounder SS et al (2009) Dietary supplementation with resveratrol reduces plaque pathology in a transgenic model of Alzheimer's disease. Neurochem Int 54(2):111–118
29. Lanz T et al (1991) The role of cysteines in polyketide synthases. Site-directed mutagenesis of resveratrol and chalcone synthases, two key enzymes in different plant-specific pathways. J Biol Chem 266(15):99716
30. Lee VM, Goedert M, Trojanowski JQ (2001) Neurodegenerative tauopathies. Annu Rev Neurosci 24:1121–1159
31. Li F et al (2012) Resveratrol, a neuroprotective supplement for Alzheimer's disease. Curr Pharm Des 18(1):27–33
32. Lublin AL, Gandy S (2010) Amyloid-beta oligomers: possible roles as key neurotoxins in Alzheimer's disease. Mt Sinai J Med 77(1):43–49
33. Madani R et al (2006) Lack of neprilysin suffices to generate murine amyloid-like deposits in the brain and behavioral deficit in vivo. J Neurosci Res 84(8):1871–1878
34. Marambaud P, Zhao H, Davies P (2005) Resveratrol promotes clearance of Alzheimer's disease amyloid-beta peptides. J Biol Chem 280(45):37377–37382
35. Markus MA, Morris BJ (2008) Resveratrol in prevention and treatment of common clinical conditions of aging. Clin Interv Aging 3(2):331–339
36. Marr RA et al (2003) Neprilysin gene transfer reduces human amyloid pathology in transgenic mice. J Neurosci 23(6):1992–1996
37. McGeer EG, McGeer PL (2001) Innate immunity in Alzheimer's disease: a model for local inflammatory reactions. Mol Interv 1(1):22–29
38. Mokuno K et al (1994) Induction of manganese superoxide dismutase by cytokines and lipopolysaccharide in cultured mouse astrocytes. J Neurochem 63(2):612–616
39. Nasri S (2012) The effect of resveratrol flavonoid on learning and memory in passive avoidance and Y maze in diabetic rats. Adv Clin Exp Med 2012, 21(6), 705–711
40. Pervaiz S, Holme AL (2009) Resveratrol: its biologic targets and functional activity. Antioxid Redox Signal 11(11):2851–2897
41. Picklo MJ et al (2002) Carbonyl toxicology and Alzheimer's disease. Toxicol Appl Pharmacol 184(3):187–197
42. Pike CJ et al (2009) Protective actions of sex steroid hormones in Alzheimer's disease. Front Neuroendocrinol 30(2):239–258

43. Ruitenberg A et al (2001) Prognosis of Alzheimer's disease: the Rotterdam Study. Neuroepidemiology 20(3):188–195
44. Sahu SS et al. (2013) Neuroprotective effect of resveratrol against prenatal stress induced cognitive impairment and possible involvement of Atpase activity. Pharmacol Biochem Behav 103(3): 520–525
45. Selkoe DJ (2000) Imaging Alzheimer's amyloid. Nat Biotechnol 18(8):823–824
46. Shaw KN, Commins S, O'Mara SM (2001) Lipopolysaccharide causes deficits in spatial learning in the watermaze but not in BDNF expression in the rat dentate gyrus. Behav Brain Res 124(1):47–54
47. Sheng JG et al (2003) Lipopolysaccharide-induced-neuroinflammation increases intracellular accumulation of amyloid precursor protein and amyloid beta peptide in APPswe transgenic mice. Neurobiol Dis 14(1):133–145
48. Simpkins JW, Wood BE, Balin BJ (2010) Alzheimer disease: the crisis is upon us. J Am Osteopath Assoc 110(9 Suppl 8):Sii-2
49. Soleas GJ, Diamandis EP, Goldberg DM (1997) Resveratrol: a molecule whose time has come? And gone? Clin Biochem 30(2):91–113
50. Sönmez U et al (2007) Neuroprotective effects of resveratrol against traumatic brain injury in immature rats. Neurosci Lett 420(2):133–137
51. Sparkman NL et al (2005) Peripheral lipopolysaccharide administration impairs two-way active avoidance conditioning in C57BL/6J mice. Physiol Behav 85(3):278–288
52. Suzuki T, Hata S (2009) [Progresses in the researches of Alzheimer's disease to reveal the molecular mechanism of disease onset and establish the biochemical diagnostics at early stage]. Nihon Ronen Igakkai Zasshi 46(5):412–415
53. Swaab DF et al (2001) Structural and functional sex differences in the human hypothalamus. Horm Behav 40(2):93–98
54. Tang MX et al (1996) Effect of oestrogen during menopause on risk and age at onset of Alzheimer's disease. Lancet 348(9025):429–432
55. Vegeto E et al (2001) Estrogen prevents the lipopolysaccharide-induced inflammatory response in microglia. J Neurosci 21(6):1809–1818
56. Wang H-W et al (2002) Soluble oligomers of beta amyloid (1-42) inhibit long-term potentiation but not long-term depression in rat dentate gyrus. Brain Res 924(2):133–140
57. Williams EJ et al (2008) The effect of Escherichia coli lipopolysaccharide and tumour necrosis factor alpha on ovarian function. Am J Reprod Immunol 60(5):462–473
58. Wood JG et al (2004) Sirtuin activators mimic caloric restriction and delay ageing in metazoans. Nature 430(7000):686–689
59. Xiao Z-M et al (2009) Estrogen regulation of the neprilysin gene through a hormone-responsive element. J Mol Neurosci 39(1–2):22–26

Chapter 13
Synthesis and Antiepileptic Activity of Schiff's Bases of Dialkylamino Alkoxy Isatin Derivatives

Konda Swathi and Manda Sarangapani

Abstract In the present work, some new 5-[2(3)-dialkylamino alkoxy] Indole 3-thiosemicarbazone 2-ones and 5-[2(3)-dialkylamino alkoxy] Indole 3-hydrazone 2-one were prepared from 5-hydroxy isatin. The structures of the products were characterized by IR, NMR, and MASS Spectral studies. All the compounds were examined for antiepileptic activity by maximal electroshock seizure (MES) and pentylenetetrazole (PTZ) induced convulsion method. These compounds were also evaluated for their neurotoxicity study by rotarod method. Some of these compounds showed good antiepileptic activity when compared with standard drug Phenytoin and all the compounds showed less neurotoxicity when compared with standard drug Diazepam.

Keywords Synthesis • 5-[2(3)-dialkylamino alkoxy] Indole 3-hydrazone 2-one • 5-[2(3)-dialkylamino alkoxy] Indole 3-thiosemicarbazone 2-ones • Antiepileptic activity

13.1 Introduction

Epilepsy, one of the common neurological disorders, is a major public health problem, affecting around 4 % of individuals over their life span. About 20–30 % of the epilepsy patients are resistant to the available medical therapies. This fact warrants the investigation for new antiepileptic drugs. Isatin is an endogenous compound isolated in 1988 and reported [1] to possess a wide range of central nervous system activities. Surendranath pandeya et al. [2] reported the synthesis and antiepileptic activity of some novel n-methyl/acetyl, 5-(un)-substituted isatin-3-thiosemicarbazones. In the last few years, isatin derivatives have been discovered which show potential hypnotic [3], antibacterial [4–6], and MAO inhibitory [7] activity.

K. Swathi (✉) • M. Sarangapani
Medicinal Chemistry Laboratory, U.C.P.Sc., Kakatiya University, Warangal 506009, Telangana, India
e-mail: kswathi84@yahoo.co.in

P. Vlamos, A. Alexiou (eds.), *GeNeDis 2014*, Advances in Experimental Medicine and Biology 822, DOI 10.1007/978-3-319-08927-0_13

It is evident from the literature survey that isatin hydrazone derivatives, isatin thiosemicarbazone derivatives, and dialkylamino alkyl derivatives showing more promising central nervous system and antiepileptic activities. Keeping in view of these two molecular moieties viz., 5-hydroxy isatin (resembles serotonin) and dialkylamino alkyl (resembles NT), it is our endeavor to bring such important moieties into a single molecular frame as a model for molecular conjunction by appropriate synthetic routes and to screen them for antiepileptic activity and neurotoxicity.

13.2 Materials and Methods

The compounds were mostly synthesized by conventional methods and described in experimental selection and also by the methods established in our laboratory.

13.2.1 Chemicals

Leptazole, Diazepam, Dialkylamino alkylhalides, Hydrazine hydrate, and Thiosemicarbazide hydrochloride were purchased from Sigma-Aldrich Chemicals Private Limited, Hyderabad, India. p-amino phenol, hydroxylamine hydrochloride, and sodium sulfate were purchased from Merck Chemicals Private Limited, Hyderabad, India.

13.2.2 Chemistry

Solvents were dried or distilled before use. Melting points were obtained on a Toshniwal melting point apparatus in open capillary tubes and are uncorrected. The purity of the compounds was ascertained by TLC on silica gel—G plates (Merck). Infrared spectra (IR) were recorded with KBR pellet on a Perkin-Elmer BX series, Infrared spectrophotometer. Mass spectra were recorded by the direct inlet method on Thadmam-mass-quantam API 400H mass spectrophotometer. ^{1}H NMR spectra were recorded on Bruker spectrospin 400 MHz spectrophotometer in DMSO-d_6. 5-hydroxy isatin was synthesized from p-amino phenol by using Sandmeyer [8] method. It consists in the reaction of aniline with chloral hydrate and hydroxylamine hydrochloride in aqueous sodium sulfate to form an isonitroso-acetanilide, which after isolation, when treated with concentrated sulfuric acid, furnishes isatin in >75 % overall yield. Melting point 210 °C.

13.2.3 *Preparation of 5-Hydroxyindole 3-Thiosemicarbazone 2-One(II) and 5-Hydroxyindole 3-Hydrazone-2-One(IV)*

5-Hydroxyisatin was heated under reflux in methanol containing two or three drops of acetic acid with thiosemicarbazide hydrochloride/hydrazine hydrate for half an hour. The product thus separated was filtered and purified by recrystalization from suitable solvent (Yield 89 %, m.p.270 °C (II), Yield 90 %, m.p.284 °C (IV)).

13.2.4 *Preparation of 5-[2(3)-Dialkylamino Alkoxy] Indole 3-Thiosemicarbazone-2-One (III) and 5-[2(3)-Dialkylamino Alkoxy] Indole 3-Hydrazone-2-One(V)*

A mixture of 5-Hydroxyindole 3-thiosemicarba-zone2-one(II)/5-hydroxyindole 3-hydrazone 2-one(IV) (0.01 mol) and dialkylamino alkylhalide (0.01 mol) placed in 10 % alcoholic potassium hydroxide and this mixture was stirred at room temperature for 6 h. The alcohol was reduced to half of its volume and cooled. The product separated was filtered, washed with small portions of cold alcohol repeatedly, and dried.

It was purified by recrystalization from hydro alcoholic mixtures to get a crystalline solid. Similarly other 5-Hydroxy isatin derivatives as shown in Fig. 13.1 were prepared and their melting points were determined in Open capillary tubes using Toshniwal melting point apparatus and are uncorrected. Purity of the compounds was checked by TLC. The physical data of the title compounds were presented in Table 13.1. The compounds were characterized by spectral data.

13.2.5 *Spectral Data*

The compounds have been characterized by the spectral data IR, PMR, and Mass.

IR spectrum (KBr) of compound **(I)** exhibited absorption bands (cm-1) 3,421.47 (OH), 1,630.08 (C=O), 1,548 (Ar, C=C), 1,282 (C–O–C), 883.85–579.8 (Ar). ^{1}**H NMR** (300 MHz, DMSO-d_6): 13.3 (s, 1H, OH), 10.36 (s, 1H, –CONH), 6.65–7.29 (m, 3 H, Ar–H). **Mass** spectrum of compound III showed molecular ion (M+) base peak at m/z (164.1).

Compound **(IIIa)** showed characteristic **IR** peaks at 3,368.41 (NH_2), 3,282.52 (CONH), 1,708 (C=O), 1,576 (Ar C=C), 1,263 (C–O), 1,085 (C=S), 1,576 (C=N), 883.85 (Ar C–C). ^{1}**H NMR** (300 MHz, DMSO-d_6): 11.36 (s, 1H, CONH), 7.29 (s, 2H, NH_2), 7.03 (s, 1H, Ar–H), 7.20 (d, 1H, Ar–H), 7.94 (d, 1H, Ar–H), 3.2 (t, 2H, O–CH_2), 2.9 (t, 2H, N–CH_2), 1.36 (s, 6H, N–$(CH_3)_2$). **Mass** spectrum of compound

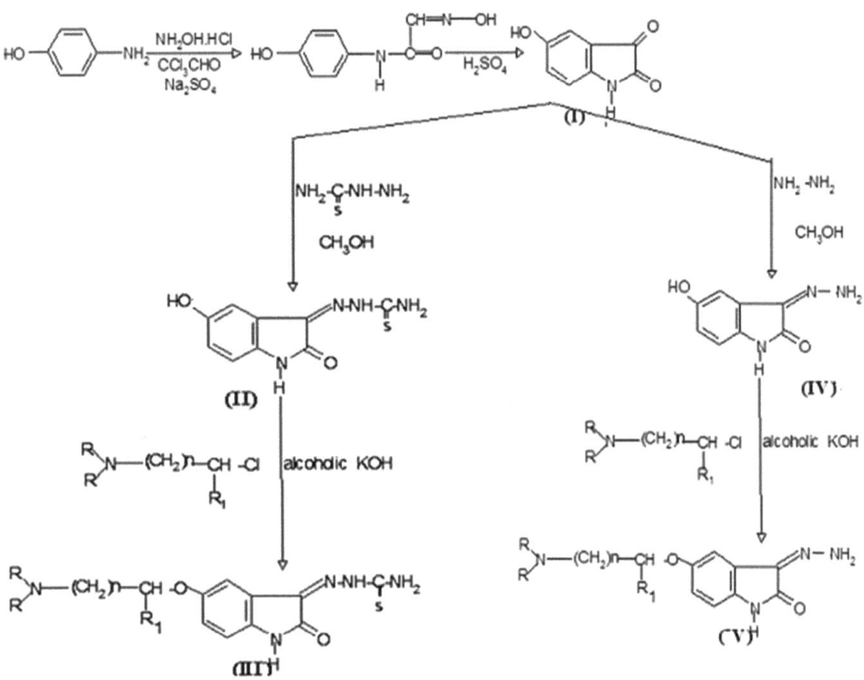

5-Hydroxy- isatin- 3- thiosemicarbazone derivatives (III)
IIIa R=CH_3; R_1=H ;n=1
IIIb R=C_2H_5; R_1=H ;n=1
IIIc R=CH_3; R_1=H ;n=2
IIId R=CH_3; R_1= CH_3 ;n=1
IIIe R=CH_3-CH-CH_3; R_1=H ;n=1

5-Hydroxy-isatin- 3-hydrazone derivatives (V)
Va R=CH_3; R_1=H ;n=1
Vb R=C_2H_5; R_1=H ;n=1
Vc R=CH_3; R_1=H ;n=2
Vd R=CH_3; R_1= CH_3 ;n=1
Ve R=CH_3-CH-CH_3;R_1=H ;n=1

Fig. 13.1 Scheme 1

IIIa showed molecular ion (M+) base peak at m/z 307. The mass spectrum shows its base peak at m/z 93 (100 %) may be due to the fragmentation of the thiosemicarbazone from the molecule ion.

Compound (**IIIb**) showed characteristic **IR** peaks at 3,368.41 (NH_2), 3,282.52 (CONH), 1,165.96 (C=S), 1,570.21 (Ar, C=C), 1,243 (C–O–C), 845.51 (Ar). **^{1}H NMR** (300 MHz, DMSO-d_6): 10.25 (s, 1H, –CONH), 7.03–7.45 (m, 3H, Ar–H), 2.99 (t, 2H, O–CH_2), 2.72 (t, 2H, N–CH_2), 7.47–7.56 (d, 2H, NH_2), 1.24 (m, 6H, N–C–CH_3), 1.12 (t, N–CH_2).

Mass spectrum of compound **IIIb** showed molecular ion (M+) base peak at m/z 335. The mass spectrum shows its base peak at m/z 214 (100 %) may be due to the fragmentation of the thiosemicarbazone from the molecule ion.

Compound (**IIIc**) showed characteristic **IR** peaks at 3,368.41 (NH_2), 3,282.52 (CONH), 1,165.96 (C=S), 1,579.72 (Ar, C=C), 1,266 (C–O–C), 805.91 (Ar). **^{1}H NMR** (300 MHz, DMSO-d_6): 10.46 (s, 1H, –CONH), 7.21–7.49 (m, 3H, Ar–H), 7.51–7.56 (d, 2H, NH_2), 2.84 (t, 2H, O–CH_2), 2.51 (m, 2H, CH_2), 2.48 (t, 2H, N–CH_2), 1.25 (s, 6H, N–$(CH_3)_2$). **Mass** spectrum of compound **IIIc** showed

Table 13.1 Physical data of 5-[2(3)-dialkylamino alkoxy] Indole 3-thiosemicarbazone-2-ones (IIIa-IIIe) and 5-[2(3)-dialkylamino alkoxy] Indole 3-hydrazone-2-ones (Va-Ve)

S.no	Compound	R	R_1	N	X	M.F	% Yield	M.P	M.Wt
1	IIIa	CH_3	H	1	$NNHCSNH_2$	$C_{13}H_{17}N_5O_2S$	91 %	280	307
2	IIIb	C_2H_5	H	1	$NNHCSNH_2$	$C_{15}H_{21}N_5O_2S$	86 %	272	335
3	IIIc	CH_3	H	2	$NNHCSNH_2$	$C_{14}H_{19}N_5O_2S$	93 %	283	353
4	IIId	CH_3	CH_3	1	$NNHCSNH_2$	$C_{14}H_{19}N_5O_2S$	85 %	264	353
5	IIIe	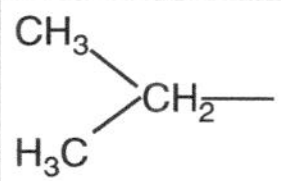	H	1	$NNHCSNH_2$	$C_{16}H_{24}N_2O_2S$	81.8 %	258	365
6	Va	CH_3	H	1	NNH_2	$C_{17}H_{27}N_5O_3$	92 %	293	248
7	Vb	C_2H_5	H	1	NNH_2	$C_{14}H_{20}N_4O_2$	83 %	269	276
8	Vc	CH_3	H	2	NNH_2	$C_{13}H_{18}N_4O_2$	92 %	261	294
9	Vd	CH_3	CH_3	1	NNH_2	$C_{13}H_{18}N_4O_2$	86 %	252	294
10	Ve	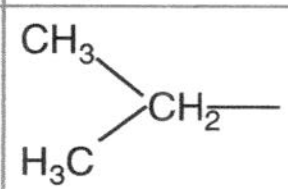	H	1	NNH_2	$C_{13}H_{26}N_4O_2$	82 %	248	306

molecular ion (M+) peak at m/z 353 (100 %). The mass spectrum shows its base peak at m/z 93 (100 %) may be due to the fragmentation of the thiosemicarbazone from the molecule ion.

Compound (**IIId**) showed characteristic **IR** peaks at 3,368.41 (NH_2), 3,282.52 (CONH), 1,165.96 (C=S), 1,546.86 (Ar, C=C), 1,245 (C–O–C), 812.71 (Ar). ^{1}**HNMR** (300 MHz, DMSO-d_6): 10.51 (s, 1H, –CONH), 7.12–7.42 (m, 3H, Ar–H), 7.51–7.56 (d, 2H, NH_2), 2.76 (m, H, O–CH), 2.45 (d, 3H, R_1=CH_3), 2.31 (d, 1H, N–CH), 1.44 (s, 6H, N–$(CH_3)_2$). **Mass** spectrum of compound **IIId** showed molecular ion (M+) base peak at m/z 353 (100 %).The mass spectrum shows its base peak at m/z 93 (100 %) may be due to the fragmentation of the thiosemicarbazone from molecule ion.

Compound (**IIIe**) showed characteristic **IR** peaks at 3,368.41 (NH_2), 3,282.52 (CONH), 1,165.96 (C=S), 1,576.34 (Ar, C=C), 1,228 (C–O–C), 814.53 (Ar). ^{1}**H NMR** (300 MHz, DMSO-d_6): 10.26 (s, 1H, –CONH), 7.34–7.51 (m, 3H, Ar–H), 7.51–7.56 (d, 2H, NH_2), 2.96 (t, 2H, O–CH_2), 2.82 (t, 2H, N–CH_2), 1.35 (t, 2H, N–CH), 1.21 (d, 12H, C–$(CH_3)_2$). **Mass** spectrum of compound **IIIe** showed molecular ion (M+) peak at m/z 365 (100 %). The mass spectrum shows its base peak at m/z 93 (100 %) may be due to the fragmentation of the thiosemicarbazone from the molecule ion.

Compound (**Va**) showed characteristic **IR** peaks at 3,450.13 (NH_2), 146.46 (CONH), 1,708 (C=O), 1,268 (C–O–C), 1,085 (C=S), 1,528 (C=N). ^{1}**H NMR** (300 MHz, DMSO-d_6): 11.36 (s, 1H, CONH), 7.29 (s, 2H, NH_2), 7.03 (s, 1H, Ar–H), 7.20 (d, 1H, Ar–H), 7.94 (d, 1H, Ar–H), 3.2 (t, 2H, O–CH_2), 2.9 (t, 2H, N–CH_2), 1.36 (s, 6H, N–$(CH_3)_2$). **Mass** spectrum of compound **Va** showed molecular ion (M+) base peak at m/z 248 (100 %). It also shows peak at m/z (71) may be due to the fragmentation of the alkyl chain from the molecule ion.

Compound (**Vb**) showed characteristic **IR** peaks at 3,450.13 (NH_2), 146.46 (CONH), 1,685.96 (C=O), 1,600.96 (C=N) 1,570.21 (Ar, C=C), 1,243 (C–O–C), 845.51 (Ar). ^{1}**HNMR** (300 MHz, DMSO-d_6): 10.25 (s, 1H, CONH), 7.03–7.45 (m, 3H, Ar–H), 2.99 (t, 2H, O–CH_2), 2.72 (t, 2H, N–CH_2), 7.47–7.56 (d, 2H, NH_2), 1.24 (s, 10H, N–$(C_2H_5)_2$).

Mass spectrum of compound **Vb** showed molecular ion (M+) peak at m/z 276 (100 %). It also shows peak at m/z (99) may be due to the fragmentation of the alkyl chain from the molecule ion.

Compound (**Vc**) showed characteristic **IR** peaks at 3,450.13 (NH_2), 146.46 (CONH), 1,698.96 (C=O), 1,600.96 (C=N), 1,579.72 (Ar, C=C), 1,266 (C–O–C), 805.91 (Ar). ^{1}**H NMR** (300 MHz, DMSO-d_6): 10.46 (s, 1H, –CONH), 7.21–7.49 (m, 3 H, Ar–H), 7.51–7.56 (d, 2H, NH_2), 2.76 (m, H, O–CH), 2.45 (d, 3H, R_1=CH_3), 2.31 (d, 1H, N–CH), 1.44 (s, 6H, N–$(CH_3)_2$). **Mass** spectrum of compound **Vc** showed molecular ion (M+) base peak at m/z 294 (100 %). It also shows peak at m/z (113) may be due to the fragmentation of the alkyl chain from the molecule ion.

Compound (**Vd**) showed characteristic **IR** peaks at 3,450.13 (NH_2), 146.46 (CONH), 1,698.96 (C=O), 1,600.96 (C=N), 1,546.86 (Ar, C=C), 1,245 (C–O–C), 812.71 (Ar). ^{1}**H NMR** (300 MHz, DMSO-d_6): 10.51 (s, 1H, –CONH), 7.12–7.42 (m, 3H, Ar–H), 7.51–7.56 (d, 2H, NH_2), 2.76 (m, 2H, O–CH_2),

2.45 (t, 3H, R_1=CH_3), 2.31 (m, 1H, N–CH), 1.44 (s, 6H, N–$(CH_3)_2$).**Mass** spectrum of compound **Vd** showed molecular ion (M+) base peak at m/z 294 (100 %). It also shows peak at m/z (113) may be due to the fragmentation of the alkyl chain from the molecule ion.

Compound (**Ve**) showed characteristic **IR** peaks at 3,450.13 (NH_2), 146.46 (CONH), 1,696.96 (C=O), 1,600.96 (C=N), 1,576.34 (Ar, C=C), 1,228 (C–O–C), 814.53 (Ar). **^{1}H NMR** (300 MHz, DMSO-d_6): 10.26 (s, 1H, –CONH), 7.34–7.51 (m, 3H, Ar–H), 7.51–7.56 (d, 2H, NH_2), 2.96 (t, 2H, O–CH_2), 2.82 (t, 2H, N–CH_2), 1.35 (m, 2H, N–CH), 1.21 (d, 12H, C–$(CH_3)_2$). **Mass** spectrum of compound **Ve** showed molecular ion (M+) base peak at m/z 306 (100 %). It also shows peak at m/z (129) may be due to the fragmentation of the alkyl chain from the molecule ion.

13.3 Pharmacology

13.3.1 Antiepileptic Activity

Materials: Normal saline, test compounds, Leptazole, stop watch, corneal electrodes, phenytoin.

Animals: Swiss mice.

13.3.1.1 Maximal Electroshock Seizure (MES) Method

Method: The antiepileptic activity was studied by Maximal Electroshock Induced Convulsion method [9] by using electro-convulsometer. Healthy male mice weighing between 20 and 25 g were fasted for overnight and divided into groups of six animals each. The test compounds suspended in normal saline were administered at a dose of 100 mg/kg body weight i.p. The control group animals received only vehicle (normal saline). The test started 30 min after i.p. injection. Maximal seizures were induced by the application of electrical current to the brain via corneal electrodes. The stimulus parameter for mice was 50 mA in a pulse of 60 Hz for 200 ms. Abolition of the hind limb tonic extensor spasm was recorded as a measure of antiepileptic activity, and results are presented in Table 13.2.

13.3.1.2 Pentylenetetrazole (PTZ) Method

Materials: Normal saline, test compounds, Leptazole, stop watch, phenytoin.

Method: The antiepileptic activity was studied by using Leptazole (Pentylenetetrazole) [10] as a chemical convulsion inducer. Healthy male mice weighing between 20 and 25 g were fasted for overnight and divided into groups of six

Table 13.2 Antiepileptic and neurotoxicity study of 5-[2(3)-dialkylamino alkoxy] Indole 3-thiosemicarbazone-2-ones (IIIa-IIIe) and 5-[2(3)-dialkylamino alkoxy] Indole 3-hydrazone-2-ones (Va-Ve)

S.no	Compound	Animals protected in % MES induced convulsions	Animals protected in % Chemically induced convulsions	Skeletal muscle relaxant activity (neurotoxicity) (%)
1	IIIa	75.54 ± 0.341	70.9 ± 1.234	6 ± 0.373
2	IIIb	68.61 ± 0.142	74.6 ± 0.341	12 ± 1.234
3	IIIc	58.42 ± 0.151	66.6 ± 0.234	6.7 ± 0.673
4	IIId	51.5 ± 1.234	59.8 ± 0.456	3.4 ± 1.095
5	IIIe	47.46 ± 0.342	54.83 ± 0.134	5.6 ± 2.384
6	Va	75.18 ± 0.436	78.2 ± 1.453	8.2 ± 1,345
7	Vb	76.68 ± 0.234	75.77 ± 1.345	3.7 ± 0.234
8	Vc	64.76 ± 0.763	54.3 ± 0.243	5.8 ± 0.567
9	Vd	48.18 ± 1.236	49.3 ± 0.653	6.8 ± 1.378
10	Ve	37.44 ± 1.451	38.9 ± 1.658	7.4 ± 1.567
11	Phenytoin	100	88.36 ± 1.234	–
12	Control	0	0	2
13	Diazepam	–	–	78 ± 0.256

Number of animals $n = 6$, $P < 0.05$
The compounds were tested at a dose of 100 mg/kg (b.w)

animals each. The animals were injected with Leptazole (80 mg/kg) given intraperitoneally. Those animals which show convulsions were selected for the experiment. The test compounds suspended in normal saline were administered at a dose of 100 mg/kg body weight i.p. The control group animals received only vehicle (normal saline). The Leptazole is again given in the same dose and the time taken for convulsions to start was noted, and results are presented in Table 13.2 and Fig. 13.2.

13.3.1.3 Neurotoxicity Study

The neurotoxicity [11] was studied by rotarod method. Healthy male mice weighing between 20 and 25 g were fasted for overnight and divided into groups of six animals each. Turn off the rotating rod, select an appropriate speed (25 rpm), and place the animal one by one on the rotating rod. A normal mouse (untreated) generally falls off within 3–5 min. Test compound dissolved in saline were administered, intraperitoneally in a dose of 100 mg/kg. The control group received saline only. One group of animals was administered diazepam as a standard (i.p 4 mg/kg). After 30 min, repeat the experiment as done earlier note the fall of time of animals before and after test compounds and diazepam treatment, respectively; results are presented in Table 13.2 and Fig. 13.3.

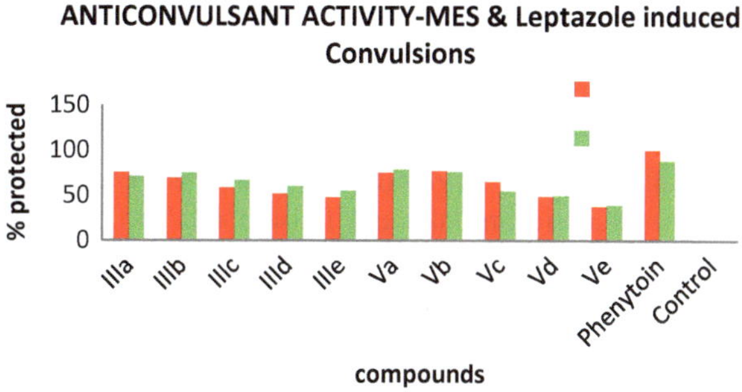

Fig. 13.2 Anticonvulsant activity of 5-[2(3)-dialkylamino alkoxy] Indole 3-thiosemicarbazone-2-ones (IIIa-IIIe) and 5-[2(3)-dialkylamino alkoxy] Indole 3-hydrazone-2-ones (Va-Ve)

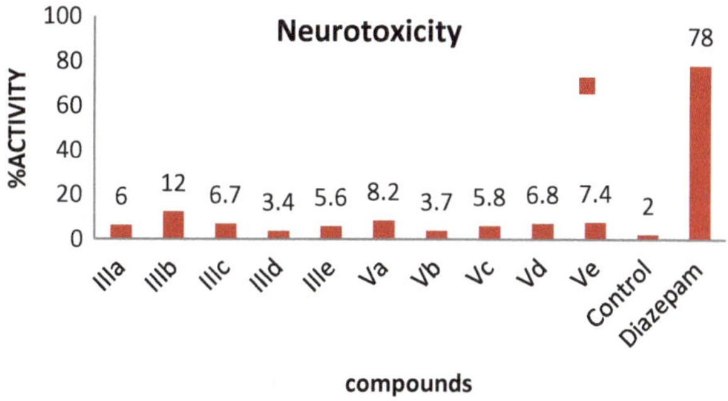

Fig. 13.3 Neurotoxicity study of 5-[2(3)-dialkylamino alkoxy] Indole 3-thiosemicarbazone-2-ones (IIIa-IIIe) and 5-[2(3)-dialkylamino alkoxy] Indole 3-hydrazone-2-ones (Va-Ve)

13.4 Results and Discussions

Physical data TLC, IR, 1H NMR, and mass spectra confirmed the structures and purity of the synthesized compounds. All the title compounds decomposed before melting. All the synthesized compounds were evaluated for their in vivo antiepileptic and skeletal muscle relaxant activity. It was observed that compounds Va, IIIa, Vb, and IIIb exhibited more promising antiepileptic activity. Among the test compounds IIIc, Vc, IIId, Vd, IIIe, and Ve were found to be next in the order of reducing the duration of convulsions. Compounds with dimethyl amino and diethyl amino ethyloxy group at C-5 of isatin showed more protection against Maximal Electroshock Seizure (MES) induced convulsions whereas compounds IIIa, IIIb and Va, Vb exhibited more protection activity against chemically (Leptazole) induced convulsions. 5-[2(3)-dialkylamino alkoxy] isatin-3-hydrazones showing more activity compared to 5-[2(3)-dialkylamino alkoxy] isatin thiosemicarbazones. All the test compounds showed less (<20 %) neurotoxicity (skeletal muscle relaxant activity) when compared with Diazepam.

Conclusion

A new series of five 5-[2(3)-dialkylamino alkoxy] Indole 2,3 dione derivatives were synthesized by reacting 5-hydroxyindole 2,3 dione schiff bases with 2-*N,N* di alkylamino alkyl halides. Evaluation of these compounds as antiepileptic and skeletal muscle relaxant activity revealed that the compounds Va($R{=}CH_3$), Vb($R{=}C_2H_5$), IIIa($R{=}CH_3$), and IIIb($R{=}C_2H_5$) with a dimethyl and diethyl amino ethyl chain derivatives were found to be relatively superior in antiepileptic activity and other compounds (IIIc, IVc, IIId, Vd, IIIe, Ve) are next in the order of activity. All the compounds showed less neurotoxicity compared to Diazepam.

Acknowledgments The first author would like to thank the CSIR, New Delhi for providing financial support. Authors are thankful to Principal University College of Pharmaceutical Sciences, Kakatiya University for providing facilities.

References

1. Bhattacharya Salil K, Mitra Shankar K, Acharya Satya B (1991) J Psychopharmacol 5:202
2. Pandeya SN, Senthil Raja A (2002) J Pharm Sci 5(3):275
3. Pandeya SN, Yogeeswari P, Stables JP (2000) Eur J Med Chem 35:879–886
4. Padhy AK, Sahu SK, Panda PK, Kar DM, Misro PK (2004) Indian J Chem 43B:971
5. Raviraj K, Ghate M, Manohar VK (2004) J Chem Sci 116(5):265
6. Gupta M, Raman S, Vikas, Srivastava SN (2004) Asian Journal of Chemistry, 16(2) 779–783
7. Gringberg B, Imazylis L, Benhena M (1990) Chemija 2:87
8. Marvel CS, Heirs GS (1941) Organ Synthesis Collect 1:327
9. Krall RL, Penry JK, White BG, Kupferberg HJ, Swinyard EA (1978) Epilepsia 19:409
10. Gerhard Vogel H (ed) (2002) Drug discovery and evaluation of pharmacological assays. 2nd edn. p 487
11. Gerhard Vogel H (ed) (2002) Drug discovery and evaluation of pharmacological assays. 2nd edn. p 398

Chapter 14
Synthesis and Screening of Biologically Significant 5-Hydroxy Isatin Derivatives for Antioxidant Activity

Konda Swathi and Manda Sarangapani

Abstract In the present work, some new 5-Hydroxyindole 3-thiosemicarbazone 2-ones and 5-[2(3)-dialkylaminoalkoxy] Indole 3-hydrazone 2-ones were prepared from 5-hydroxy isatin. The structures of the products were characterized by IR, NMR, and MASS Spectral studies. All the compounds were examined for antioxidant activity by using 1,1-diphenyl-2-picryl-hydrazyl and hydrogen peroxide (H_2O_2), and the total antioxidant capacity by a phosphomolybdenum assay. In general, the derivatives were found to exhibit antioxidant activity. Further, the compounds with dialkyl amino alkoxy at the C5 position demonstrated significant antioxidant activity.

Keywords Synthesis • 5-[2(3)-Dialkyl amino alkoxy] Indole 3-hydrazone 2-one • 5-Hydroxyindole 3-thiosemicarbazone 2-ones • Antioxidant activity

14.1 Introduction

Isatin is an endogenous compound isolated in 1988 and reported [1] to possess a wide range of central nervous system activities. Surendranath pandeya et al. [2] reported the synthesis and anticonvulsant activity of some novel n-methyl/acetyl, 5-(un)-substituted isatin-3-thiosemicarbazones. In the last few years, Isatin derivatives have been discovered which show potential hypnotic [3], antibacterial [4–6], and MAO inhibitory [7] activity.

It is evident from the literature survey that Isatin derivatives, isatin thiosemicarbazone derivatives, and dialkylamino alkyl derivatives showing more promising antioxidant activities. Keeping in view of these two molecular moieties viz., 5-hydroxy isatin (Resembles serotonin) and dialkyl-amino alkyl (Resembles NT), it is our endeavor to bring such important moieties into a single molecular frame as

K. Swathi (✉) • M. Sarangapani
Medicinal Chemistry Laboratory, U.C.P.Sc., Kakatiya University,
Warangal 506009, Telangana, India
e-mail: kswathi84@yahoo.co.in; panimanda@gmail.com

P. Vlamos, A. Alexiou (eds.), *GeNeDis 2014*, Advances in Experimental
Medicine and Biology 822, DOI 10.1007/978-3-319-08927-0_14

a model for molecular conjunction by appropriate synthetic routes and to screen them for antioxidant activity.

14.2 Materials and Methods

The compounds were mostly synthesized by conventional methods and described in experimental selection and also by the methods established in our laboratory.

14.2.1 Chemicals

Dialkylaminoalkylhalides, Hydrazinehydrate, and Thiosemicarbazide hydrochloride were purchased from Sigma-Aldrich Chemicals Private Limited, Hyderabad, India. p-amino phenol, hydroxylamine hydrochloride, and sodium sulfate were purchased from Merck Chemicals Private Limited, Hyderabad, India. 1,1-Diphenyl-2-picryl-hydrazyl (DPPH) and hydrogen peroxide (H_2O_2) were obtained from Merck (India).

14.2.2 Chemistry

Solvents were dried or distilled before use. Melting points were obtained on a Toshniwal melting point apparatus in open capillary tubes and are uncorrected. The purity of the compounds was ascertained by TLC on silica gel–G plates (Merck). Infrared spectra (IR) were recorded with KBR pellet on a Perkin-Elmer BX series, Infrared spectrophotometer. Mass spectra were recorded by the direct inlet method on Thadmam-mass-quantam API 400H mass spectrophotometer. ^{1}H NMR spectra were recorded on Brucker spectrospin 400 MHz spectrophotometer in DMSO-d6. 5-hydroxy Isatin was synthesized from p-amino phenol by using Sandmeyer [8] method. It consists in the reaction of aniline with chloral hydrate and hydroxylamine hydrochloride in aqueous sodium sulfate to form an isonitrosoacetanilide, which after isolation, when treated with concentrated sulfuric acid, furnishes isatin in >75 % overall yield.

14.2.3 Preparation of 5-Hydroxyindole 3-Thiosemicarbazone 2-One(II) and 5-Hydroxyindole 3-Hydrazone-2-One(IV)

5-Hydroxyisatin was heated under reflux in methanol containing two or three drops of acetic acid with thiosemicarbazide hydrochloride/Hydrazine hydrate for half an hour. The product thus separated was filtered and purified by recrystalization from suitable solvent (Yield 89 %, m.p. 270 °C(II), Yield 90 %, m.p. 284(IV)).

14.2.4 Preparation of 5-[2(3)-Dialkyl Amino Alkoxy] Indole 3-Hydrazone-2-One(III) and 5-[2(3)-Dialkyl Amino Alkoxy] Indole 3-Thiosemicarbazone-2-One(V)

A mixture of 5-hydroxy indole 3-hydrazone 2-one(IV)/5-Hydroxy in-dole3-thiosemicarbazone2-one(II) (0.01 mol) and dialkylamino alkylhalide (0.01 mol) placed in 10 % alcoholic potassium hydroxide and this mixture was stirred at room temperature for 6 h. The alcohol was reduced to half of its volume and cooled. The product separated was filtered, washed with small portions of cold alcohol repeatedly, and dried. It was purified by recrystalization from hydro alcoholic mixtures to get a crystalline solid. Similarly other 5-Hydroxy Isatin derivatives as shown in Fig. 14.1 were prepared and their melting points were determined in Open capillary tubes using Toshniwal melting point apparatus and are uncorrected. Purity of the compounds was checked by TLC. The physical data of the title compounds were presented in Table 14.1. The compounds were characterized by spectral data.

14.2.5 Spectral Data

The compounds have been characterized by the spectral data IR, PMR, and Mass.

IR spectrum (KBr) of compound (I) exhibited absorption bands (cm^{-1}) 3,421.47 (OH), 1,630.08(C=O), 1,548(Ar,C=C), 1,282(C–O–C), 883.85–579.8(Ar). ^1H NMR (300 MHz, DMSO-d6): 13.3(s, 1H, OH), 10.36(s,1H,-CONH), 6.65–7.29 (m,3H,Ar–H). Mass spectrum of compound III showed molecular ion (M+) base peak at *m/z* (164.1).

Compound (IIIa) showed characteristic IR peaks at 3,368.41(NH_2), 3,282.52 (CONH), 1,708(C=O), 1,576(Ar C=C), 1,263(C–O), 1,085(C=S), 1,576(C=N), 883.85(Ar, C–C). ^1H NMR (300 MHz,DMSO-d6): 11.36(s,1H,CONH), 7.29(s,2H, NH_2), 7.03(s,1H,Ar–H), 7.20(d,1H,Ar–H), 7.94(d,1H,Ar–H), 3.2(t,2H,O–CH_2), 2.9 (t,2H,N–CH_2), 1.36(s, 6H,N-$(CH_3)_2$). Mass spectrum of compound IIIa showed molecular ion (M+) base peak at *m/z* 307. The mass spectrum shows its base peak at

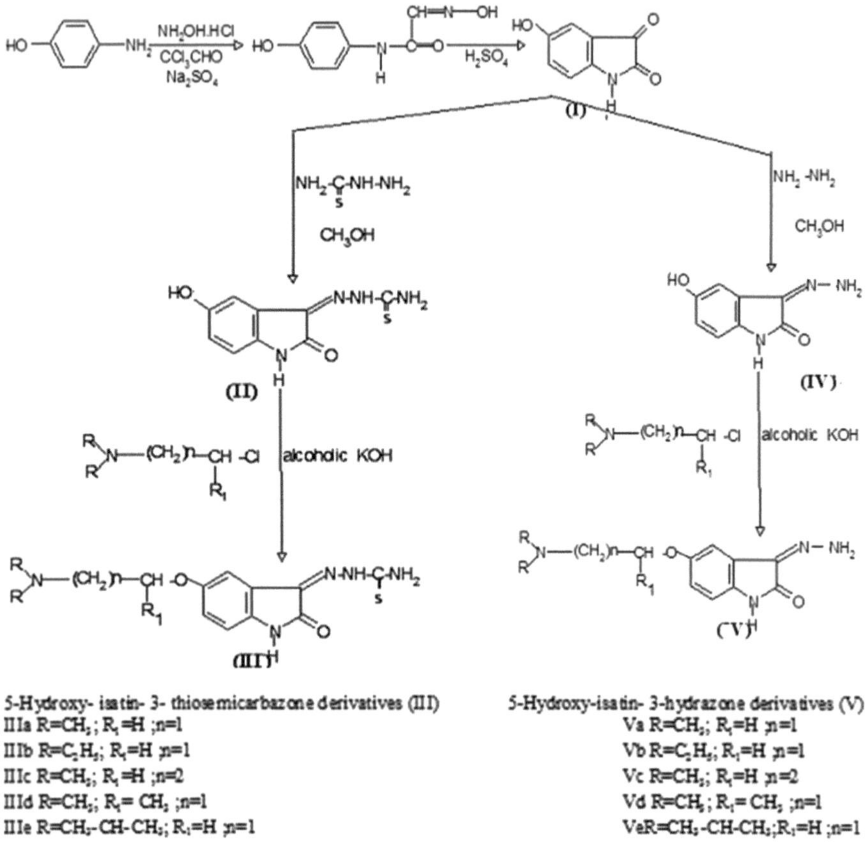

Fig. 14.1 Scheme 1

Table 14.1 Physical data of 5-[2(3)-dialkyl amino alkoxy] Indole 3-thiosemicarbazone-2-ones (IIIa-IIIe) and5-[2(3)-dialkyl amino alkoxy] Indole 3-hydrazone-2-ones(Va-Ve)

S. no	Compound	R	R_1	N	X	M.F	% Yield	M.P	M.Wt
1	IIIa	CH_3	H	1	$NNHCSNH_2$	$C_{13}H_{17}N_5O_2S$	91 %	280	307
2	IIIb	C_2H_5	H	1	$NNHCSNH_2$	$C_{15}H_{21}N_5O_2S$	86 %	272	335
3	IIIc	CH_3	H	2	$NNHCSNH_2$	$C_{14}H_{19}N_5O_2S$	93 %	283	353
4	IIId	CH_3	CH_3	1	$NNHCSNH_2$	$C_{14}H_{19}N_5O_2S$	85 %	264	353
5	IIIe	CH_3 CH_2— H_3C	H	1	$NNHCSNH_2$	$C_{16}H_{24}N_2O_2S$	81.8 %	258	365
6	Va	CH_3	H	1	NNH_2	$C_{17}H_{27}N_5O_3$	92 %	293	248
7	Vb	C_2H_5	H	1	NNH_2	$C_{14}H_{20}N_4O_2$	83 %	269	276
8	Vc	CH_3	H	2	NNH_2	$C_{13}H_{18}N_4O_2$	92 %	261	294
9	Vd	CH_3	CH_3	1	NNH_2	$C_{13}H_{18}N_4O_2$	86 %	252	294
10	Ve	CH_3 CH_2— H_3C	H	1	NNH_2	$C_{13}H_{26}N_4O_2$	82 %	248	306

m/*z* 93 (100 %) may be due to the fragmentation of the thiosemicarbazone from the molecule ion.

Compound (IIIb) showed characteristic IR peaks at 3368.41(NH_2), 3,282.52 (CONH), 1,165.96 (C=S), 1,570.21 (Ar ,C=C), 1,243(C–O–C), 845.51(Ar). ^{1}H NMR (300 MHz, DMSO-d6): 10.25(s, 1H,-CONH), 7.03-7.45(m,3H,Ar–H), 2.99 (t,2H,O–CH_2), 2.72(t,2H,N–CH_2), 7.47-7.56(d,2H,NH_2), 1.24(m,6H,N–C–CH_3,1.12(t,N–CH_2).

Mass spectrum of compound IIIb showed molecular ion (M+) base peak at *m*/*z* 335. The mass spectrum shows its base peak at *m*/*z* 214 (100 %) may be due to the fragmentation of the thiosemicarbazone from the molecule ion.

Compound (IIIc) showed characteristic IR peaks at 3,368.41(NH_2), 3,282.52 (CONH), 1,165.96(C=S), 1,579.72 (Ar,C=C), 1,266(C–O–C), 805.91(Ar). ^{1}H NMR (300 MHz, DMSO-d6): 10.46(s,1H,-CONH), 7.21-7.49(m,3 H,Ar–H), 7.51-7.56(d,2H, NH_2), 2.84(t,2H,O–CH_2), 2.51(m,2H,CH_2), 2.48(t,2H,N–CH_2), 1.25 (S,6H,N-$(CH_3)_2$). Mass spectrum of compound IIIc showed molecular ion (M+) peak at *m*/*z* 353 (100 %). The mass spectrum shows its base peak at *m*/*z* 93 (100 %) may be due to the fragmentation of the thiosemicarbazone from the molecule ion.

Compound (IIId) showed characteristic IR peaks at 3,368.41(NH_2), 3,282.52 (CONH), 1,165.96(C=S), 1,546.86 (Ar,C=C), 1,245(C–O–C), 812.71(Ar). ^{1}H NMR (300 MHz,DMSO-d6): 10.51(s,1H,-CONH), 7.12-7.42(m,3H,Ar–H), 7.51-7.56(d,2H,NH_2), 2.76(m,H,O–CH), 2.45(d,3H,R1=CH_3), 2.31(d,1H,N–CH), 1.44 (s,6H,N-$(CH_3)_2$). Mass spectrum of compound IIId showed molecular ion (M+) base peak at *m*/*z* 353(100 %). The mass spectrum shows its base peak at *m*/*z* 93 (100 %) may be due to the fragmentation of the thiosemicarbazone from molecule ion.

Compound (IIIe) showed characteristic IR peaks at 3,368.41(NH_2), 3,282.52 (CONH), 1,165.96(C=S), 1,576.34 (Ar,C=C), 1,228(C–O–C), 814.53(Ar). ^{1}H NMR (300 MHz, DMSO-d6): 10.26(s, 1H,-CONH), 7.34-7.51(m,3H,Ar–H), 7.51–7.56(d,2H,NH_2), 2.96(t,2H,O–CH_2), 2.82(t,2H,N–CH_2), 1.35(t, 2H,N–CH), 1.21(d,12H,C -$(CH_3)_2$). Mass spectrum of compound IIIe showed molecular ion (M +) peak at *m*/*z* 365 (100 %). The mass spectrum shows its base peak at *m*/*z* 93 (100 %) may be due to the fragmentation of the thiosemicarbazone from the molecule ion.

Compound (Va) showed characteristic IR peaks at 3450.13(NH_2), 146.46 (CONH), 1,708(C=O), 1,268(C–O–C), 1,085(C=S), 1,528(C=N). ^{1}H NMR (300 MHz, DMSO-d6): 11.36(s,1H,CONH), 7.29(s,2H,NH_2), 7.03(s,1H,Ar–H), 7.20(d,1H,Ar–H), 7.94(d,1H,Ar–H), 3.2(t,2H,O–CH_2), 2.9(t,2H,N–CH_2), 1.36 (s,6H,N-$(CH_3)_2$). Mass spectrum of compound Va showed molecular ion (M+) base peak at *m*/*z* 248 (100 %). It also shows peak at *m*/*z* (71) may be due to the fragmentation of the alkyl chain from the molecule ion.

Compound (Vb) showed characteristic IR peaks at 3,450.13(NH_2), 146.46 (CONH), 1,685.96(C=O), 1,600.96(C=N), 1,570.21(Ar,C=C), 1,243(C–O–C), 845.51(Ar). ^{1}H NMR (300 MHz,DMSO-d6): 10.25(s,1H,CONH), 7.03-7.45 (m,3H,Ar–H), 2.99(t,2H,O–CH_2), 2.72(t,2H,N–CH_2), 7.47-7.56(d,2H,NH_2), 1.24 (s,10H,N-$(C_2H_5)_2$).

Mass spectrum of compound Vb showed molecular ion (M+) peak at *m/z* 276 (100 %). It also shows peak at *m/z* (99) may be due to the fragmentation of the alkyl chain from the molecule ion.

Compound (Vc) showed characteristic IR peaks at 3,450.13(NH_2), 146.46 (CONH), 1,698.96(C=O), 1,600.96(C=N), 1,579.72(Ar, C=C), 1,266(C–O–C), 805.91(Ar). ^{1}H NMR (300 MHz,DMSO-d6): 10.46(s,1H,-CONH), 7.21–7.49 (m,3H,Ar–H), 7.51–7.56(d,2H, NH_2), 2.76(m,H,O–CH), 2.45(d,3H, R1=CH_3), 2.31(d,1H,N–CH), 1.44(s,6H,N-$(CH_3)_2$). Mass spectrum of compound Vc showed molecular ion (M+) base peak at *m/z* 294 (100 %). It also shows peak at *m/z* (113) may be due to the fragmentation of the alkyl chain from the molecule ion.

Compound (Vd) showed characteristic IR peaks at 3450.13(NH_2), 146.46 (CONH), 1,698.96(C=O), 1,600.96(C=N), 1,546.86(Ar,C=C), 1,245(C–O–C), 812.71(Ar). ^{1}H NMR (300 MHz,DMSO-d6): 10.51(s,1H,-CONH), 7.12–7.42 (m,3H,Ar–H), 7.51–7.56(d,2H,NH_2), 2.76(m,2H,O–CH_2), 2.45(t,3H,R1=CH_3), 2.31(m,1H,N–CH), 1.44(s,6H,N-$(CH_3)_2$). Mass spectrum of compound Vd showed molecular ion (M+) base peak at *m/z* 294 (100 %). It also shows peak at *m/z* (113) may be due to the fragmentation of the alkyl chain from the molecule ion.

Compound (Ve) showed characteristic IR peaks at 3,450.13(NH_2), 146.46 (CONH), 1,696.96 (C=O), 1,600.96 (C=N), 1,576.34(Ar,C=C), 1,228(C–O–C), 814.53(Ar). ^{1}H NMR (300 MHz,DMSO-d6): 10.26(s,1H,-CONH), 7.34–7.51 (m,3H,Ar–H), 7.51–7.56(d,2H,NH_2), 2.96(t,2H,O–CH_2), 2.82(t,2H,N–CH_2), 1.35 (m,2H,N–CH), 1.21(d,12H,C-$(CH_3)_2$). Mass spectrum of compound Ve showed molecular ion (M+) base peak at *m/z* 306(100 %). It also shows peak at *m/z* (129) may be due to the fragmentation of the alkyl chain from the molecule ion.

14.3 Pharmacology

14.3.1 DPPH Radical Scavenging Method

Blios [9] showed that an α,α-diphenyl-β-picryl hydrazyl radical can be used for determining antioxidant activity. DPPH in ethanol shows a strong absorption band at 517 nm (independent of pH from 5 to 6.5), and the solution appears to be deep violet in color. As the DPPH radical is scavenged by the donated hydrogen from the antioxidant, the absorbance is diminished according to the stoichiometry. Briefly, 0.5 ml of DPPH solution (0.2 mM) was mixed with 0.1 ml of various concentrations of test compounds and 1.5 ml ethanol added. The mixture was kept at room temperature for 30 min, and then absorbance (OD) was read at 517 nm against blank. The % reduction of free radical concentration (OD) with different concentrations of test compounds was calculated and compared with the standard, ascorbic acid. The results were expressed as IC50 values (the concentration of test required to scavenge 50 % free radicals).

14.3.2 *Hydrogen Peroxide Scavenging Activity*

The ability of test compounds to scavenge hydrogen peroxide was determined by using the method of Sanchez (2001) and Famey et al. [10]. The solution of hydrogen peroxide (20 mM) was prepared in phosphate buffered saline (pH 7.4). Various concentrations of 1 ml of test compounds and standards were added to 2 ml of H_2O_2. Absorbance of hydrogen peroxide at 230 nm was determined 10 min later against the blank. Ascorbic acid was used as a reference standard.

14.3.3 *Total Antioxidant Capacity by a Phosphomolybdenum Assay*

The in vitro antioxidant activity of the synthesized compounds was evaluated by the phosphomolybdenum method according to the procedure of Prieto, Pineda, and Aguilar [11]. The principle of the assay is based on the reduction of Mo (VI) to Mo (V) by test compounds and subsequent formation of a green phosphate/Mo(V) complex at acid pH.

An aliquot of 0.1 ml of the test solution in methanol was mixed with 1 ml of a reagent solution (0.6 M sulfuric acid, 28 mM sodium phosphate, and 4 mM ammonium molybdate).

The tubes were capped and incubated at 950 °C for 90 min. The samples were cooled at room temperature and then absorbance was measured at 695 nm against the blank. The blank solution was containing 1 ml of the reagent solution and an appropriate volume of the same solvent used in the test compound. The total antioxidant capacity of the tested compounds was calculated according to the equation

$$\text{TAC}\ (\%) = (\text{A0} - \text{At}/\text{A0}) \times 100$$

where At is the absorbance value of the test compound and A0 is the absorbance of the blank sample. The reference standard is ascorbic acid.

14.4 Results and Discussions

Physical data TLC, IR, ^{1}H NMR, and mass spectra confirmed the structures and purity of the synthesized compounds. All the title compounds decomposed before melting. All the compounds were examined for antioxidant activity by using 1,1-diphenyl-2-picryl-hydrazyl and hydrogen peroxide (H_2O_2), and the total antioxidant capacity by a phosphomolybdenum assay.

Table 14.2 In vitro antioxidant activity of DPPH and hydrogen peroxide, total antioxidant capacity by the phosphomolybdenum assay of synthesized isatin derivatives

S.no	Compounds	IC50 (μg/ml) DPPH	IC50 (μg/ml) H_2O_2	TAC (μg/ml)
1	IIIa	55.54 ± 0.041	50.9 ± 1.234	22.96 ± 0.373
2	IIIb	48.61 ± 0.042	54.6 ± 0.341	212 ± 1.234
3	IIIc	38.42 ± 0.051	36.6 ± 0.234	346.7 ± 0.673
4	IIId	31.5 ± 1.204	29.8 ± 0.456	413.4 ± 1.095
5	IIIe	27.46 ± 0.042	24.8 ± 0.134	525.7 ± 2.384
6	Va	55.18 ± 0.036	58.2 ± 1.453	128.2 ± 1,345
7	Vb	56.68 ± 0.034	55.7 ± 1.345	233.7 ± 0.234
8	Vc	44.76 ± 0.063	38.9 ± 0.243	345.8 ± 0.567
9	Vd	28.18 ± 1.036	24.3 ± 0.243	523.8 ± 1.378
10	Ve	27.44 ± 1.0451	29.3 ± 0.653	537.4 ± 1.567
11	Ascorbic acid	16.55 ± 0.063	21.96 ± 0.243	–

Values are expressed in mean ± SD, $N = 3$; TAC = total antioxidant capacity

DPPH radical scavenging activity
The interaction of the synthesized compounds IIIa–Ve with the stable free radical DPPH is presented in Table 14.2. The values are expressed in IC50, i.e., the ability of the test compound required to decrease the concentration of free radicals by 50 %. The IC50 values of the test compounds were found between 27 and 55.54 mg/ml. Free radical scavenging of synthesized compounds rose with increase in concentration. Compounds IIIe, Vd, and Ve with IC50 values 27.46 mg/ml, 28.18 mg/ml, and 27.4 mg/ml, respectively, demonstrated potent antioxidant activity compared to other compounds and the standard, ascorbic acid. The results of this study revealed that all the compounds are significantly scavenged by the DPPH free radical.

Hydrogen peroxide scavenging activity
H_2O_2 scavenging power is based upon the ability of a compound to convert H_2O_2 into water and was used for the determination of the H_2O_2 scavenging power. The measurement of H_2O_2 scavenging activity may be one of the useful methods for determining the ability of antioxidants to decrease the level of pro-oxidants such as H_2O_2. Hydrogen peroxide is a weak oxidizing agent capable of oxidizing the essential thiol (–SH) groups of proteins, thus inactivating a few enzymes. H_2O_2 readily penetrates cell membranes and inside the cell reacts with Fe^{2+} to form a hydroxyl radical which exerts several adverse effects. The combination of reduced iron and hydrogen peroxide gives a hydroxyl radical in the well-known Fenton reaction.

$$Fe^{2+} + H_2O_2 \rightarrow Fe^{3+} + OH^- + OH$$

The radical scavenging effect of the synthesized compounds against H_2O_2 was measured spectrophotometrically. Values for their IC50 are given in Table 14.2. Compounds IIId, Ve, IIIe, and Ve were found to have potent antioxidant activity with IC50 below 50, i.e., 29.8 mg/ml, 29.3 mg/ml, 24.8 mg/ml, and 24.3 mg/ml, respectively. It was observed that compounds Va, IIIa, Vb, and IIIb exhibited less promising antioxidant activity. Among the test compounds IIIc and Vc were found to show moderate activity.

Determination of the total antioxidant capacity
The antioxidant activity of the synthesized compounds was evaluated by the phosphomolybdenum method. The assay is based on the reduction of Mo(VI) to Mo(V) by the test compounds and subsequent formation of a green phosphate/Mo(V) complex at acid pH. TAC ranges from 22.96 to 537.4 mg/ml. The most active compound was Ve. In addition, experimental data showed compounds Ve, IIId, and Vd possessed significant TAC. The phosphomolybdenum method is an alternative to methods already available for the evaluation of TAC due to its simplicity and the cheap reagents it uses. It is quantitative, since the antioxidant activity is expressed as the number of equivalents of ascorbic acid.

Conclusion
A new series of five 5-[2(3)-dialkyl amino alkoxy] Indole 2,3 dione derivatives were synthesized by reacting 5-hydroxyindole 2,3 dione schiff bases with 2-*N,N* dialkylamino alkyl halides. Evaluation of these compounds as antioxidant activity revealed that the compounds Vd, Ve, IIId, and IIIe with a diisopropyl and diethyl amino ethyl/propyl chain derivatives were found to be relatively superior in antioxidant activity and other compounds (IIIc, Vc, IIIa, Va, IIIb, Vb) are next in the order of activity.

Acknowledgments The first author would like to thank the CSIR, New Delhi for providing financial support. Authors are thankful to Principal University College of Pharmaceutical Sciences, Kakatiya University for providing facilities.

References

1. Bhattacharya Salil K, Mitra Shankar K, Acharya Satya B (1991) J Psychopharmacol 5:202
2. Pandeya SN, Senthil Raja A (2002) J Pharm Sci 5(3):275
3. Pandeya SN, Yogeeswari P, Stables JP (2000) Eur J Med Chem 35:879–886
4. Padhy AK, Sahu SK, Panda PK, Kar DM, Misro PK (2004) Indian J Chem 43B:971
5. Kusanur RA, Ghate M, Kulkarni MV (2004) J Chem Sci 116(5):265
6. Gupta S, Raman, Vikas SN, Srivastava (2004) Asian J Chem 16(2):779–783
7. Gringberg B, Imazylis L, Benhena M (1990) Chemija 2:87
8. Marvel CS, Heirs GS (1941) Organ Synth Collect 1:327
9. Blios MS (1958) Antioxidant determination by the use of stable free radical. Nature 26:1199–1200
10. Famey EJC, Luyengi L, Lee SK, Zhu LF, Zhou BN, Prog HHS (1998) Antioxidant flavanoid glycosides from Daphniphyllum calycium. J Nat Prod 61:706–708
11. Prieto P, Pineda M, Aguilar M (1999) Spectrophotometric quantitation of antioxidant capacity through the formation of phosphomolybdenum complex: specific application to the determination of vitamin E1. Anal Biochem 269:337–341

Chapter 15
Modeling Neural Circuits in Parkinson's Disease

Maria Psiha and Panayiotis Vlamos

Abstract Parkinson's disease (PD) is caused by abnormal neural activity of the basal ganglia which are connected to the cerebral cortex in the brain surface through complex neural circuits. For a better understanding of the pathophysiological mechanisms of PD, it is important to identify the underlying PD neural circuits, and to pinpoint the precise nature of the crucial aberrations in these circuits. In this paper, the general architecture of a hybrid Multilayer Perceptron (MLP) network for modeling the neural circuits in PD is presented. The main idea of the proposed approach is to divide the parkinsonian neural circuitry system into three discrete subsystems: the external stimuli subsystem, the life-threatening events subsystem, and the basal ganglia subsystem. The proposed model, which includes the key roles of brain neural circuit in PD, is based on both feed-back and feed-forward neural networks. Specifically, a three-layer MLP neural network with feedback in the second layer was designed. The feedback in the second layer of this model simulates the dopamine modulatory effect of compacta on striatum.

Keywords Parkinson's disease • Human brain neural activity • Basal ganglia • Multilayer perceptron (MLP) neural network

15.1 Introduction

Parkinson's disease (PD) is just one of the several neurologic movement disorders that produce similar symptoms. In certain cases patients rapidly become totally disabled; in others, the disease progresses extremely slowly; and in yet others, illness is chronic (always present) and may have more severe symptoms as time goes on. Due to the fact that the physical properties, or evolution, of these diseases vary greatly, proper diagnosis is crucial. Progressive resting tremor (4–7 Hz), rigidity, bradykinesia/akinesia, gait disturbance, and postural instability characterize PD clinically [1]. The degeneration or death of dopamine-generating cells in the pars compacta region of the substantia nigra leads to alterations in the activity of the

M. Psiha (✉) • P. Vlamos
Department of Informatics, BiHELab, Ionian University, Plateia Tsirigoti 7, 49100 Corfu, Greece
e-mail: psiha@ionio.gr; vlamos@ionio.gr

P. Vlamos, A. Alexiou (eds.), *GeNeDis 2014*, Advances in Experimental Medicine and Biology 822, DOI 10.1007/978-3-319-08927-0_15

neural circuits within the basal ganglia that regulate movement resulting in a difficulty to perform both voluntary and involuntary movements. Research has failed to determine why these brain cells begin to die or why they continue to die.

A variety of research approaches have so far been considered to study and model neural activity in PD including electrophysiological recordings of neuronal activity, biochemical analyses, images studies, and computational modeling. Specifically, recent progress in neuroscience research has led to major insights into the structure and function of the basal ganglia and into the pathophysiologic basis of PD. Models of basal ganglia circuitry have provided piece of evidence into the pathophysiological mechanisms of movement disorders and reasons that disruption of parts of the circuit by lesion or deep brain stimulation might improve motor function. In the literature animal models of Parkinson's disease, such as the 6-hydroxydopamine (6-OHDA) rat model and the 1-methyl-4-phenyl-1,2,3,6-tetrahydropyridine (MPTP) monkey model, have been proposed for the understanding of neurophysiological changes that give rise to the pathophysiology of the disease [2–5]. These changes include increases in firing rate, a tendency to fire in a more irregular pattern, and abnormal oscillatory synchronization. In addition, the renaissance of stereotactic surgery for PD has provided valuable neuronal recording and imaging data from human subjects. Newer genetic models, for instance mice that overexpress synuclein, provide further insights into the genetic backdrop upon which PD develops.

In this paper, we present the general architecture of a hybrid Multilayer Perceptron (MLP) network for modeling the neural circuits in PD. The main idea of the proposed approach is to divide the parkinsonian neural circuitry system into three discrete subsystems: the external stimuli subsystem, the life-threatening events subsystem, and the basal ganglia subsystem. In accordance with basal ganglia structure, the proposed model uses both feed-back and feed-forward neural networks. Specifically, a three-layer MLP neural network with feedback in the second layer was designed. The feedback in the second layer of this model simulates the dopamine modulatory effect of compacta on striatum. The proposed model will be trained with human brain electrophysiological data from normal persons and PD patients in order to elucidate the changes in brain neural circuit

15.2 Physiological Background

No specific, standard criteria exist for the neuropathologic diagnosis of Parkinson disease, as the specificity and sensitivity of its characteristic findings have not been clearly established. Parkinson's symptoms are most commonly treated pharmacologically, with variants of L-dopa, a chemical precursor of the neurotransmitter dopamine. Before the introduction of levodopa, Parkinson disease caused severe disability or death in 25 % of patients within 5 years of onset, 65 % within 10 years, and 89 % within 15 years. The mortality rate from Parkinson disease was three times that of the general population matched for age, sex, and racial origin. With the introduction of levodopa, the mortality rate dropped approximately 50 %, and longevity was extended by many years. L-dopa treatment can be quite effective for overcoming akinesia, but it typically has at least two undesirable characteristics:

1. Its effects are widely variable throughout the course of a dose (the "on/off" response).
2. It typically results in increasingly more severe dyskinesia (excessive movement) over time, often to the point of being as difficult to manage as the presenting symptom (akinesia) and often resulting in profound weight loss due to constant exertion.

Growing evidence in the PD literature suggests that the neurodegeneration of the dopamine-producing cells in the basal ganglia impairs the automated and internally paced mode of movement control, but spares the functionality of goal-directed actions [6, 7]. Basal ganglia (BG) input is from cortex and BG output is relayed through thalamus to supplementary motor area (SMA). Based on neuronal structure and neurotransmitters released, BG is supposed to be composed of five neuronal blocks (Fig. 15.1): Substantia Nigra (SN), Globus Pallidus (GP), Subthalamic Nucleus (STN), Putamen, and Caudate. Putamen and caudate, which are separated by the internal capsule, act as BG inputs and are called collectively "striatum" (Str).

GP is divided into the internal segment (GPi) and external segment (GPe). Both divisions of the GP have distinctive functionality and differ from their inputs and outputs. In a similar way, the SN is divided into two functionally different parts: Compacta (SNc) and Pars Reticulata (SNr). SNc and SNr have the same input but have different outputs and have neurochemical distinct neuron types. SNc mainly consist of dopamine-generating cells.

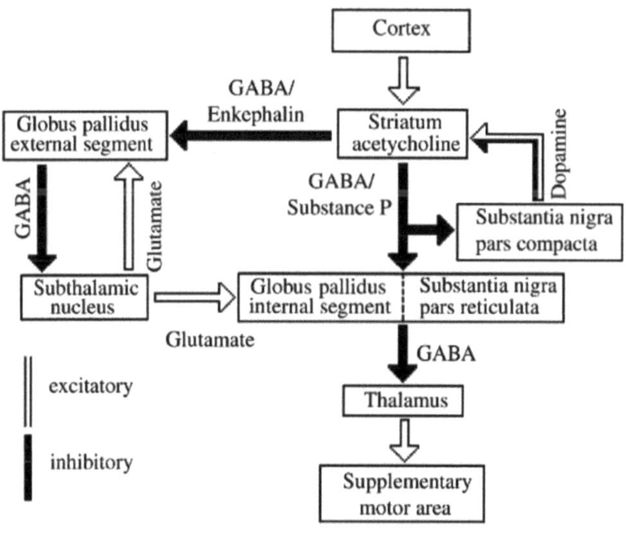

Fig. 15.1 Conceptual model of BG (data obtained from [14])

In 1989 and 1990 researchers proposed the famous "box-and-arrow" model of BG connectivity and physiology so-called the direct pathway and indirect pathway and is still the most influential and accepted contribution [8, 9]. The first connects Str to BG output and the second connects Str to BG output through STN and GPe. These two pathways are controlled by dopamine signal via the modulatory effect of SNc. There are no direct outputs from the BG to spinal or brainstem motor neurons. The BG influence motor, sensory, and cognitive cortical information through the thalamus [10]. Head and eye movements are also influenced through the superior colliculus [11], and spinal cord processing and locomotion and postural control features through pedunculopontine nucleus (PPN) [12]. When ~90% of cells producing dopamine in SNc are impaired, symptoms of PD appear [13].

The canonical model of basal ganglia dysfunction proposes that alterations in neuronal firing rates underlie the spectrum of movement disorders. In Parkinson's disease, there is decreased activity of the inhibitory direct pathway connecting the basal ganglia striatal input to its output in the GPi and increased activity in the so-called indirect pathway via the external globus pallidum (GPe) and STN.

15.2.1 Paradoxical Kinesia

Paradoxical kinesia (PK) is the sudden transient ability of a patient with Parkinson's disease (PD) to perform a task he or she was previously unable to perform [15]. This phenomenon has been puzzling scientists for over 60 years, both in neurological and motor control research, with the underpinning mechanism still being the subject of fierce debate. The sensory cues governing this behavior and the prevalence in real-life situations are unknown. The neural theory of paradoxical kinesia assumes that the action triggered by external stimuli bypasses degenerated basal ganglia pathways [16]. In the literature many studies have been proposed for the occurrence of paradoxical kinesia in PD [17]. PK has also been considered as a general property of the motor system, as it has been observed in other diseases like autism and Asperger's disease [18].

According to a model proposed by Redgrave [6], PK can be accounted for by preserved goal-directed loop functionality taking over impaired circuitry of self-generated habitual control. The basal ganglia operate on the basis of micro-circuitries that share a unique somatotopic and behavioral specialization in processing different sources of sensory information. This information is produced by reactions to an external cue or a life-threatening event. The term "cue" can be explained as a stimulus presented in space and time that provides information about movement initiation and/or execution [19].

It has long been observed that patients with PD can sometimes react and move quickly in response to external visual and auditory stimuli. Experimental results indicate that the auditory stimuli act as pacemaker function and provide an external rate which is able to stabilize the defective rate within the basal ganglia [20]. Previous efforts to use sound to manage gait impairments in PD patients have typically involved rhythmic auditory stimulation [21]. This type of intervention generally provides PD patients with an auditory metronome, or markedly rhythmic music, and asks them to match consecutive footfalls with the onset of each beat. In spite of some success with this approach in reducing gait variability, other research suggests that the technique of using metronomic stimuli is limited in its effect [22]. Researchers also underline the benefits of dancing in the normalization of motor activity in PD patients [23].

In the literature many visual cues have been proposed for the stimulation of paradoxical kinesia [24]. For these visual cues there are a variety of factors that must be considered. Researchers tried to identify the kind of visual cues (static or dynamic) required for normal gait [24]. Visual cues are believed to activate specific motor pathways in the brain. These pathways allow the damaged circuits to be bypassed, resulting in normal motor function. Cerebellar sensory-motor pathways are currently being explored to try and prove this hypothesis [25].

The sudden resolution of a previously stabilized akinesia in a PD patient has been also reported in the case of facing an immediate threat. Most cases cited have occurred when an immediate threat was posed to the patient, such as that of a runaway horse, an oncoming car, a fire, an earthquake, or a war [26–29]. In these situations, an otherwise immobile patient was suddenly able to flee. For obvious ethical reasons, researchers cannot create a situation stressful enough to test the occurrence of paradoxical kinesia during life-threatening events.

15.3 Overall Structure of the Proposed Model

The proposed system is divided into the following three subsystems (Fig. 15.2):

1. External Stimuli Subsystem
2. Life-Threatening Events Subsystem
3. Basal Ganglia Subsystem

The entries of the proposed system are signals from the External Stimuli Subsystem and Life-Threatening Events Subsystem. The assumed architecture includes a feed-back network with one layer (Elman network) and a Multilayer Perceptron (MLP) feed-forward network with three layers. Specifically, SNc is modeled with an Elman network. The remaining parts of the basal ganglia, the external stimuli subsystem, and the life-threatening events subsystem are modeled with an MLP neural network. Multilayer networks solve the classification problem for nonlinear sets by employing hidden layers, whose neurons are not directly connected to the output. The additional hidden layers can be interpreted geometrically as additional hyperplanes, which enhance the separation capacity of the network.

As we mentioned before, the balance between output of the direct and indirect pathway is controlled by a difference between actions of dopamine, from SNc, on striatal neurons. Physiological findings show a feed-back structure for dopamine

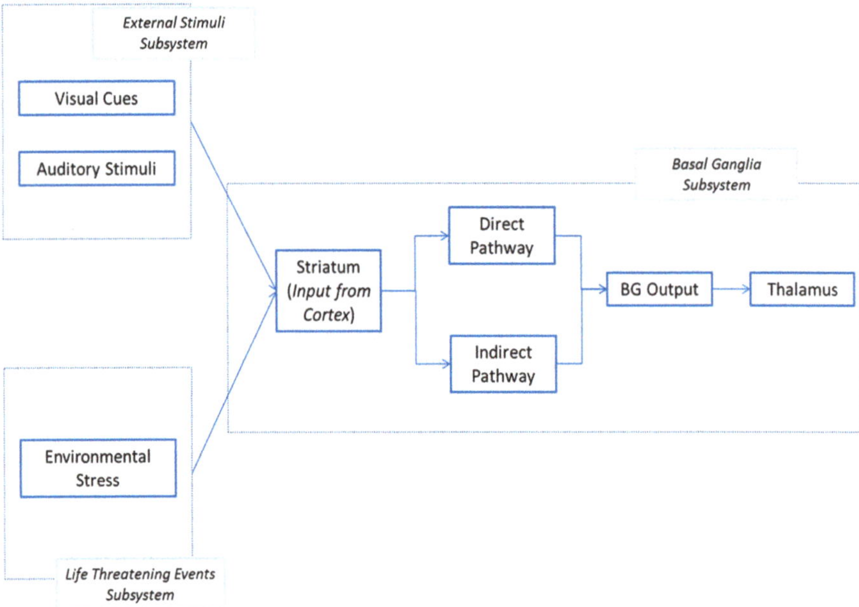

Fig. 15.2 The general structure of the proposed system

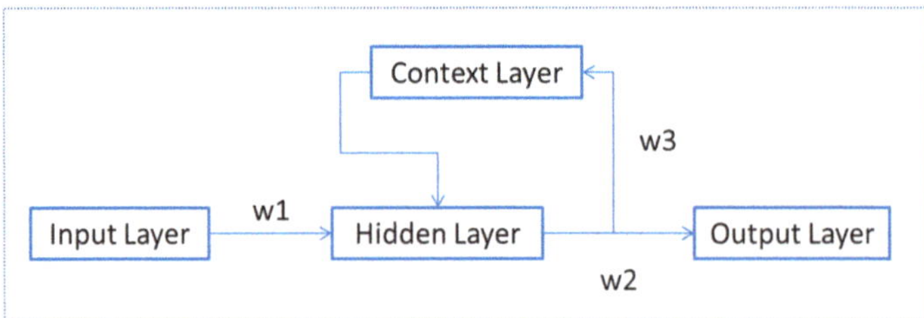

Fig. 15.3 The structure of an Elman's recurrent neural network

modulatory effect in basal ganglia that has a critical role in PD [30]. For modeling this part of BG we use a Simple Recurrent Network (SRN) or Elman Network which is an MLP with a single hidden layer and it contains connections from the hidden layer's neurons to the context units. The context units store the output values from the hidden neurons in a time unit and these values are fed as additional inputs to the hidden neurons in the next time unit [31]. Elman Network can be easily trained by the back-propagation algorithm.

The structure of an Elman's Recurrent Neural Network is illustrated in Fig. 15.3.

Weight matrices are as follows: W1 is the weight matrix between input layer and hidden layer, W3 the weight matrix between context layer and hidden layer, and

W2 is the weight matrix between hidden layer and output layer. The outputs of the neurons in the hidden layer and output layer can be computed by:

$$y_j^s = f\left(\sum_{i=1}^{n} w1_{ij}x_i^s + \sum_{j=1}^{m} w3_{ij}y_j^{s-1}\right)$$

$$z_k^s = f\left(\sum_{j=1}^{m} w2_{jk}y_j^s\right)$$

where f is the activation function of each neuron [32].

In feed-forward networks the flow of data is strictly from input to output cells that can be grouped into layers but no feed-back interconnections can exist. If we put W3 = 0 in Fig. 15.3, we can find the schema of feed-forward structure.

Conclusion

The brain structure is so complex that it is impossible to recognize its performance in different states exactly. In PD, this complexity is usually increased. Although pathological basal ganglia activity has been well characterized in studies of parkinsonian patients or animal models of the disease, the dynamics of the modifications undergone by basal ganglia neuronal activity during the evolution period of the disease are not known yet.

Computational models present a global understanding of a specific brain function in PD and normal states. There are different computational models for PD, most of which are black boxes that have considered some details of physiological findings. The gait behavior and its disorders are complex and for this reason, designing a model based on physiological findings may be useful in understanding the disease and finally to diagnose and control it in earlier stages [30]. In this paper, we present a model which includes the key roles of brain neural circuit in PD. The proposed model will be trained with human brain electrophysiological data from normal persons and PD patients in order to elucidate the changes in brain neural circuit. The current models of basal ganglia pathophysiology are incomplete and should be taken as a first draft of basal ganglia dysfunction in the different disease states. Most pertinently, changes in phasic discharge patterns, and new anatomic connections need to be better incorporated into any new concept of basal ganglia function and a greater emphasis placed on the manner in which thalamic, brainstem, and cortical neurons utilize basal ganglia output.

References

1. Jankovic J (2008) Parkinson's disease: clinical features and diagnosis. J Neurol Neurosurg Psychiatry 79:368–376
2. Raz A, Feingold A, Zelanskaya V, Vaadia E, Bergman H (1996) Neuronal synchronization of tonically active neurons in the striatum of normal and parkinsonian primates. J Neurophysiol 76:2083–2088
3. Raz A, Vaadia E, Bergman H (2000) Firing patterns and correlations of spontaneous discharge of pallidal neurons in the normal and the tremulous 1-methyl-4-phenyl-1,2,3,6-tetrahydropyridine vervet model of parkinsonism. J Neurosci 20:8559–8571
4. Ruskin DN, Bergstrom DA, Tierney PL, Walters JR (2003) Correlated multisecond oscillations in firing rate in the basal ganglia: modulation by dopamine and the subthalamic nucleus. Neuroscience 117:427–438
5. Goldberg JA, Rokni U, Boraud T, Vaadia E, Bergman H (2004) Spike synchronization in the cortex-basal ganglia networks of parkinsonian primate reflects global dynamics of the local field potentials. J Neurosci 24:6003–6010
6. Redgrave P, Rodriguez M, Smith Y, Rodriguez-Oroz M, Lehericy S, Bergman H, Agid Y, DeLong MR, Obeso JA (2010) Goal-directed and habitual control in the basalganglia: implications for Parkinson's disease. Nat Rev Neurosci 11:760–772
7. Torres EB, Heilman KM, Poizner H (2011) Impaired endogenously evoked auto-mated reaching in Parkinson's disease. J Neurosci 31:17848–17863
8. Albin R, Young A, Penney J (1989) The functional anatomy of basal ganglia disorders. Trends Neurosci 12:366–375
9. DeLong M (1990) Primate models of movement disorders of basal ganglia origin. Trends Neurosci 13:281–285
10. Kandel E, Schwartz J, Jessel T (2000) Principles of neural science. McGraw-Hill, New York, NY
11. Hikosaka O, Takikawa Y, Kawagoe R (2000) Role of the basal ganglia in the control of purposive saccadic eye movements. Physiol Rev 80:953–978
12. Takakusaki K, Oohinata-Sugimoto J, Saitoh K, Habaguchi T (2004) Role of basal gangliabrainstem systems in the control of postural muscle tone and locomotion. Prog Brain Res 143:231–237
13. Marsden C (1990) Parkinson's disease. Lancet 335:948–592
14. Guyton AC, Hall J (2001) Text book of medical physiology, 10th edn. Saunders, Philadelphia, PA
15. Glickstein M, Stein J (1991) Paradoxical movement in Parkinson's disease. Trends Neurosci 14(11):480–482
16. Martin J (1967) Disorder of locomotion associated with disease of the basal ganglia. In: The Basal Ganglia and Posture. Philadelphia: Lippincott, 24–35
17. Snijders AH, Bloem BR (2010) Cycling for freezing of gait. N Engl J Med 362(13):e46
18. Rinehart N, McGinley J (2010) Is motor dysfunction core to autism spectrum disorder? Dev Med Child Neurol 52(8):697
19. Horstink M, De Swart B, Wolters E, Berger H (1993) Paradoxical behavior in Parkinson'sdisease. In: Wolters EC, Scheltens P (eds) Mental dysfunction in Parkinson's disease. Proceedings of the European congress on mental dysfunction in Parkinson's disease. Vrije Universiteit, Amsterdam
20. Salimpoor VN et al (2011) Anatomically distinct dopamine release during anticipation and experience of peak emotion to music. Nat Neurosci 14(2):257–262
21. Hausdorff JM, Lowenthal J, Herman T, Gruendlinger L, Peretz C, Giladi N (2007) Rhythmic auditory stimulation modulates gait variability in Parkinson's disease. Eur J Neurosci 26 (8):2369–2375
22. Wittwer JE, Webster KE, Hill K (2013) Music and metronome cues produce different effects on gait spatiotemporal measures but not gait variability in healthy older adults. Gait and Posture 37(2):219–222

23. Houston S, McGill A (2013) A mixed-methods study into ballet for people living with Parkinson's. Arts Health 5(2):103–119
24. Azulay J-P et al (1999) Visual control of locomotion in Parkinson's disease. Brain 122(1):111–120
25. Ferrarin M, Brambilla M, Garavello L, Di Candia A, Pedotti A, Rabuffetti M (2004) Microprocessor-controlled optical stimulating device to improve the gait of patients with Parkinson's disease. Med Biol Eng Comput 42(3):328–332
26. Daroff RB (2008) Paradoxical kinesia. Mov Disord 23(8):1193–1193
27. Hammond TC (2010) New developments: falls, drooling & exercise in Parkinson's disease. The Parkinson's source. Issue 40. APDA magazine
28. Bonanni L et al (2010) Protracted benefit from paradoxical kinesia in typical and atypical parkinsonisms. Neurol Sci 31(6):751–756
29. Schlesinger I, Erikh I, Yarnitsky D (2007) Paradoxical kinesia at war. Mov Disord 22 (16):2394–2397
30. Sarbaz Y, Gharibzadeh S, Towhidkhah F, Banaie M, Jafari A (2011) A gray-box neural network model of Parkinson's disease using gait signal. Basic Clin Neurosci 2(3)
31. Haykin S (2008) Neural networks and learning machines, 3rd edn. Pearson Education, Upper Saddle River, NJ
32. Seker S, Ayaz E, Turkcan E (2003) Elman's recurrent neural network applications to condition monitoring in nuclear power plant and rotating machinery. Eng Appl Artif Intel 16:647–656

Chapter 16
Neuropathology of Parkinsonism in Alzheimer's Disease

Eniko Kovari, Constantin Bouras, François R. Herrmann, Pierre R. Burkhard, Judit Horvath, and Aikaterini Xekardaki

Keywords Parkinsonism • Alzheimer's disease • Neuropathology • Substantia nigra • Basal ganglia

About 30 % of Alzheimer's disease (AD) patients develop parkinsonian features, but Lewy body pathology is not always present at autopsy. So the neuropathological substrate of extrapyramidal signs in AD remains unclear. In the present study neuronal and neurofibrillary tangle (NFT) densities were counted in the substantia nigra pars compacta (SN) and in the putamen of 22 AD patients, 11 with and 11 without parkinsonism. Parkinsonism was defined as the presence of bradykinesia and at least one of resting tremor, rigidity, or gait disorders. Our results showed that parkinsonism in AD is related to a significant neuronal loss both in the SN and in the putamen, suggesting pre- and postsynaptic alterations of the nigrostriatal pathway.

In contrast, neuronal tau deposition was a less important factor. Only densities of NFTs in the SN correlated with parkinsonism but not in the putamen. We propose that a subgroup of pure AD patients develop parkinsonian symptoms as a result of neuronal loss in the basal ganglia, indicating a prominent subcortical involvement, which appears unrelated to the neuropathological severity of AD.

E. Kovari (✉) • C. Bouras • A. Xekardaki
Department of Psychiatry, University Hospitals, Geneva, Switzerland
e-mail: eniko.kovari@hcuge.ch; Constantin.Bouras@unige.ch; Aikaterini.Xekardaki@hcuge.ch

F.R. Herrmann
Department of Internal Medicine, Rehabilitation and Geriatrics, University Hospitals, Geneva, Switzerland
e-mail: Francois.Herrmann@hcuge.ch

P.R. Burkhard • J. Horvath
Department of Neurology, University Hospitals, Geneva, Switzerland
e-mail: Pierre.Burkhard@hcuge.ch; judit.Horvath@hcuge.ch

P. Vlamos, A. Alexiou (eds.), *GeNeDis 2014*, Advances in Experimental Medicine and Biology 822, DOI 10.1007/978-3-319-08927-0_16

Chapter 17
Towards an Expert System for Accurate Diagnosis and Progress Monitoring of Parkinson's Disease

Athanasios Alexiou, Maria Psiha, and Panayiotis Vlamos

Abstract While Parkinson's disease is a chronic and progressive movement disorder, no one can predict which symptoms will affect an individual patient. At the present time there is no cure for Parkinson's disease but instead a variety of alternative treatments provide relief from the symptoms. Due to these unpromising factors, we propose a new multi-scale ontology-based modeling technology for the accurate diagnosis of Parkinson's disease and its progress monitoring. The proposed model will be used to assess the status of the patient with PD corresponding treatments using a multilayer neural network. The proposed tool also aims to identify new associated physical and biological biomarkers from heterogeneous patients' data. The architecture of this expert system and its implementation in Protégé is presented in this paper.

Keywords Expert system • Parkinson's disease • Ontology-based modeling • Multilayer neural network

17.1 Introduction

Parkinson's disease (PD) is a neurodegenerative brain disorder and belongs to a group of conditions called motor system disorders, which are the result of the loss of dopamine-producing brain cells. Worldwide, it is estimated that four to six million people suffer from PD.[1] It is difficult to estimate how quickly or slowly PD will progress in each patient. Due to the fact that the physical properties, or evolution, of the disease varies greatly, proper diagnosis and accurate progress monitoring is crucial.

[1] http://www.parkinson.org/parkinson-s-disease.aspx

A. Alexiou • M. Psiha (✉) • P. Vlamos
BiHELab, Department of Informatics, Ionian University,
Plateia Tsirigoti 7, 49100 Corfu, Greece
e-mail: alexiou@ionio.gr; psiha@ionio.gr; vlamos@ionio.gr

P. Vlamos, A. Alexiou (eds.), *GeNeDis 2014*, Advances in Experimental Medicine and Biology 822, DOI 10.1007/978-3-319-08927-0_17

While the pathological causes of PD have not been specified yet, today there is no single test or biomarker that can predict whether a particular person will develop PD and a definitive diagnosis is only possible after death—with postmortem analysis [1]. The brain changes that create neurodegenerative diseases such as PD are microscopic, on a chemical level and non-blood tests or electroencephalograms (EEGs) and MRI/CAT scans can definitively diagnose PD.

With no specific diagnostic tools, physicians must prove their scientific diagnosis for PD based on the physical examination of the patient and some general symptoms. Physicians are intimately familiar with the characteristic history and the signs and symptoms found when examining a person with Parkinson's. Then, they have to judge how closely the history of symptoms and the neurologic findings of any specific person match those of typical Parkinson's disease.

The molecular pathogenesis of PD is still not understood, while at least two forms of PD exist: idiopathic (sporadic) and heritable (familiar) [2]. In familiar PD linkage analysis has identified a number of PD-associated genes, as well as disruption of mitochondrial dynamics [3–8]. Specifically, two genes Pink1 and Parkin have been identified in hereditary PD, both of which have been shown to be important for mitochondrial integrity [8]. Interpretation of Pink1 and Parkin functions have been complicated by the discrepancies in mitochondrial morphology defects, found among various mammalian cell lines and between the fly and mammalian model systems [7]. Additionally, after years of intense studies, a considerable number of scientific researches demonstrated the important role of mitochondrial dysfunction and oxidative stress to development of the more common neurogenerative diseases, like Alzheimer's disease, Parkinson's disease, and Huntington's disease [9]. Until now, although the role of mitochondrial dysfunctions in PD has been identified [4–7], it is not clear how and why these dysfunctions cause PD symptoms.

Considering the above, we present the general architecture of a hybrid ontology-based expert system for the accurate diagnosis and progress monitoring of PD. The proposed system manages to identify physical, biological, environmental, and genetic risk factors that are underlying causes of PD. The proposed knowledge base has been developed in Protégé platform, defining the relations between our data. For the assessment of the patient's status with the appropriate treatment we design an expert system using a multilayer neural network. The proposed hybrid expert system is a new efficient decision support software that physicians can use to diagnose and to monitor the progress of PD in clinical cases using heterogeneous patients' data. Latest studies demonstrate that a patient's quality of life deteriorates quickly if treatment is not instituted at or shortly after diagnosis [10]. The main criterion for the evaluation is the diagnostic accuracy. In addition, the cost-effectiveness of various combinations of biomarkers in PD diagnostics should be studied. This information will be used to optimize diagnostic protocols.

The proposed approach is based on the complete model that has been proposed in [4] for the description of dysfunctions in the mitochondrial membrane and it's denoted as the phenomenon of "electric thrombosis." According to this phenomenon, electric complexes result in higher concentration of protons H+, leading to the interruption of the normal flow of electrons, resulting in the inadequate production of ATP. This phenomenon affects the energy requirements of the cell and the required electrical capacity of mitochondrial membranes.

17.2 Materials and Methods

17.2.1 Treatment and Progress Monitoring

As we have already mentioned, there is no cure for PD, but instead a variety of treatments provide relief from the symptoms. Usually, patients are given levodopa combined with *carbidopa*. According to the U.S. Food and Drug Administration (FDA) and NIH, *carbidopa* delays the conversion of levodopa into dopamine until it reaches the brain and nerve cells can use levodopa to make dopamine and replenish the brain's dwindling supply.

Before the introduction of levodopa, PD caused severe disability or death in 25 % of patients within 5 years of onset, 65 % within 10 years, and 89 % within 15 years. The mortality rate from PD was three times more of the general population matched for age, sex, and racial origin. With the introduction of levodopa, the mortality rate dropped approximately 50 %, and longevity was extended by many years. This is thought to be due to the symptomatic effects of levodopa, as no clear evidence suggests that levodopa stems the progressive nature of the disease [10, 12]. Additionally, other medicines such as *bromocriptine*, *pramipexole*, and *ropinirole* mimic the role of dopamine in the brain, causing the neurons to react as they would to dopamine and may help control tremor and rigidity. An antiviral drug, *amantadine*, also appears to reduce symptoms. In May 2006, the FDA approved *rasagiline* to be used along with levodopa for patients with advanced PD or as a single-drug treatment for early PD.

In some cases, surgery may be appropriate if the disease doesn't respond to drugs. A therapy called deep brain stimulation (DBS) has now been approved by the FDA. In DBS, electrodes are implanted into the brain and connected to a small electrical device called a pulse generator that can be externally programmed. DBS can reduce the need for levodopa and related drugs, which in turn decreases the involuntary movements called dyskinesia that are a common side effect of levodopa. It also helps to alleviate fluctuations of symptoms and to reduce tremors, slowness of movements, and gait problems.

The American Academy of Neurology notes that the following clinical features may help predict the rate of progression of Parkinson's disease [13]:

- Older age at onset and initial rigidity/hypokinesia can be used to predict:
  - A more rapid rate of motor progression in those with newly diagnosed Parkinson's disease

  - Earlier development of cognitive decline and dementia; however, initially presenting with tremor may predict a more benign disease course and longer therapeutic benefit from levodopa

- A faster rate of motor progression may also be predicted if the patient is male, has associated comorbidities, and has postural instability/gait difficulty (PIGD)
- Older age at onset, dementia, and decreased responsiveness to dopaminergic therapy may predict earlier nursing home placement and decreased survival

Furthermore, PD is no longer considered only as a movement disorder; it is a neurophysiological and neuropsychiatric disorder with many non-motor symptoms that can be as or more debilitating than the motor symptoms [14]. Developing in parallel with this broadening view of the disease is the recognition that we need to widen our therapeutic arsenal by not only focusing on medical treatment of motor symptoms but also including nonmedical therapies. Exercise is one of the most useful non-medical treatments for patients with PD. Exercise can improve strength and balance, help with weight management, alleviate depression, and improve quality of life [15].

Aspects of the patient history that suggest PD include gradual onset and steady progression. The United Kingdom Parkinson Disease Society Brain Bank (UKPDS BB) has published Diagnostic Criteria that are widely accepted and used for diagnosing PD in clinical practice and for research purposes. According to the UKPDS BB criteria for PD, signs must include bradykinesia and at least one of the following: muscular rigidity, rest tremor, or postural instability. In addition, patients should not have any atypical features, such as early memory loss, early hallucinations, severe dysautonomia, cerebellar signs, or supranuclear gaze palsy. Finally, a list of supportive criteria that increase the probability of idiopathic PD is included, such as asymmetric onset, maintained asymmetry over time, and response to levodopa [16].

17.2.2 The General Structure of the Proposed System

The general structure of the proposed system is described below (Fig. 17.1):

The entries of the proposed system are data from the patient's health record. Patient's health record consists of a range of data, including PD progress monitoring data, PD associated genes, and patient's demographic data. PD progress monitoring data include:

- Neuropsychological test scores
- Brain tissue samples
- Genetic material (DNA and RNA) and their mutations
- Molecular data (metabolomics and proteomics)
- Imaging data (MRI imaging, PET (FDG/PIB) imaging)
- Electrophysiological data (TMS/EEG)

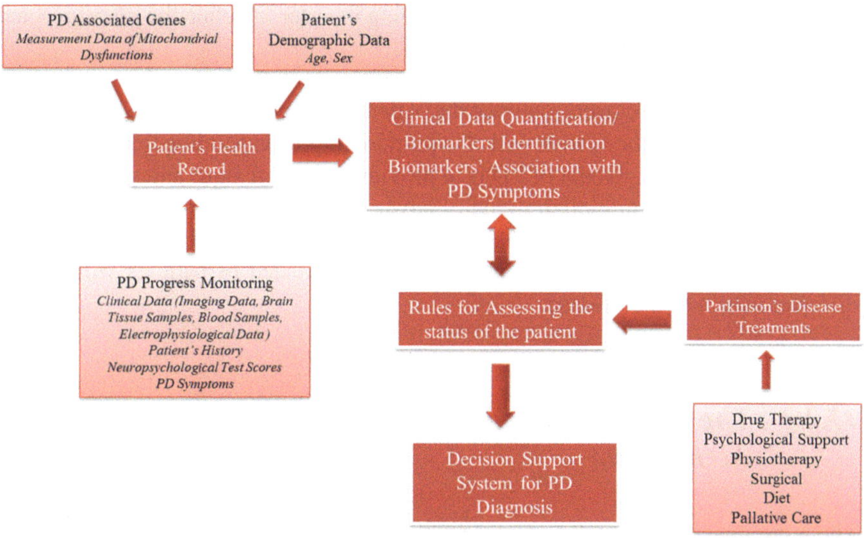

Fig. 17.1 The general structure of the proposed system

- Biomarkers detected from blood (metabolomics, proteomic)
- Measurement data of mitochondrial dysfunctions (such as mitochondrial DNA, four mitochondrial complexes, pH and ATP levels, and energy potential in the inner mitochondrial membrane)

The proposed model also represents the association of these heterogeneous patient data and tries to extract various biomarkers. Then the identified from the above level biomarkers are associated with PD symptoms. PD should be regarded as a multi-regional, multi-system neurodegenerative disorder in which the pathology appears in a regionally specific sequence [15]. The classic motor features of PD typically start insidiously and emerge slowly over weeks or months, with tremor being the most common initial symptom [17]. The three cardinal signs of PD are resting tremor, rigidity, and bradykinesia. Postural instability (balance impairment) is sometimes listed as the fourth cardinal feature. However, balance impairment in PD is a late phenomenon and in fact, prominent balance impairment in the first few years suggests that PD is not the correct diagnosis. Non-motor symptoms can be categorized as autonomic, cognitive/psychiatric and sensory [18] and may include depression, dementia, hallucinations, rapid eye movement (REM), sleep behavior disorder (RMD), orthostatic hypotension, and constipation. Non-motor symptoms can also fluctuate, especially depression, pain, numbness, paresthesia, akathisia, and restless-legs syndrome. Recognition of non-motor symptoms of Parkinson's disease is essential for appropriate management [18].

At the next level the system aims to assess the status of the patient with PD corresponding treatments using a multilayer neural network. This network will use supervised learning techniques for data training in order to define the relevance and efficiency of sets of biomarkers.

Finally, a decision support system will be developed, predicting if the patient has the PD and if yes, the current status of the patient. Depending on the stage of disease (early or advanced), the system will also propose the corresponding treatments. The goal of medical management of PD is to provide control of signs and symptoms for as long as possible while minimizing of adverse effects is occurred.

For the validation of the proposed model researchers can use retrospective patient data and can also create patient groups consisted of patients with PD in different stages of the disease.

17.2.3 The Ontology-Based Data Source Integration Architecture

The more general approach to representation of deep knowledge in expert systems is provided by hybrid systems [19]. In the proposed system, the knowledge base has been designed using *ontology-based modeling*, which combines structured vocabularies with a small set of relationships between their components. Specifically, we used a shared vocabulary of terms for describing gene product characteristics, gene product annotation and diseases, as well as tools to access and process this data by using biological ontologies. The specification of a vocabulary includes definitions of classes, relations, functions, and other objects.

In our hybrid ontology-based data source integration architecture, the ontologies and the data sources are represented as classes and semantic relationships between data source content and ontology are represented as links (Fig. 17.2). The main components of this hybrid ontology method are:

1. Using a shared ontology
2. Building the main ontology consisted of the main class and the local subclasses. Classes constitute a taxonomic hierarchy
3. Defining the roles or properties. Roles describe the properties or attributes of our classes
4. Defining the facets that specify the properties (constraints) of classes' roles
5. Creating the individual instances of classes with specific values for our roles

Specifically, we create the main ontology "Parkinson's Disease Ontology" that will be used as an extension of the shared ontology "Disease ontology."[2] The main class of the proposed system named "Patient_Care" will be used for progress monitoring of Parkinson's disease. It also identifies the current status of the patient and the corresponding treatments are given, depending on the stage of disease

[2] http://disease-ontology.org/

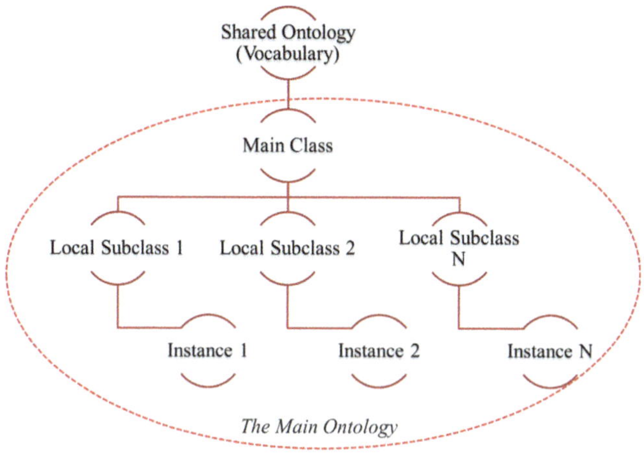

Fig. 17.2 The hybrid ontology-based data source integration architecture

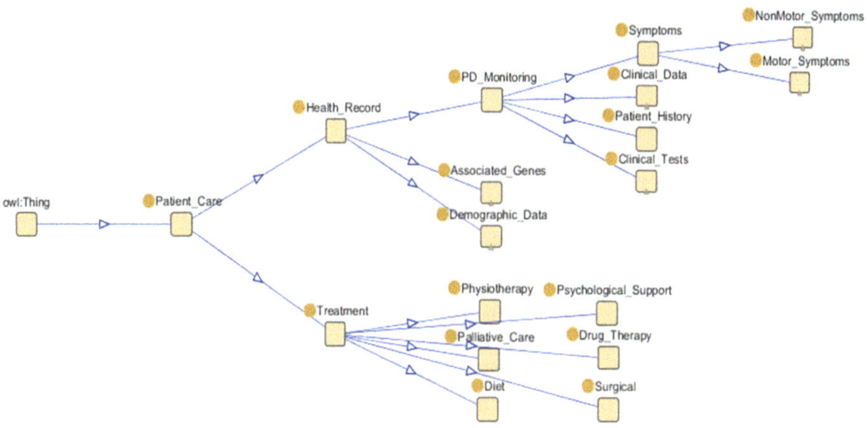

Fig. 17.3 The graphical representation of the classes

(early or advanced). Some of the local subclasses of the system are the following (Fig. 17.3):

1. Health_Record
   (a) PD_Monitoring
      - Clinical_Data
      - Clinical_Tests
      - Symptoms

        Motor_Symptoms
        NonMotor_Symptoms
      - Patient_History

(b) Demographic_Data

- Age
- Sex

(c) Associated_Genes

- ATP13A2
- DJ-1
- LRRK2
- PARK
- Parkin
- Pink1

2. Treatment

(a) Drug Therapy
(b) Psychological_Support
(c) Physiotherapy
(d) Surgical
(e) Diet
(f) Palliative_Care

The graphical representation of these classes is a digraph called "knowledge tree" (Fig. 17.3).

In the literature a few ontology-based models have been proposed for modeling PD [20]. Well-known examples of biological ontologies also include the *Gene Ontology Project*[3] and *BioPAX*.[4] Such ontologies can be used to provide unique biological identification for each component in the proposed PD model.

Open-source ontology editors and knowledge-based frameworks for knowledge representation in PD can be also used. In this paper, we used Protégé Version 3.4.8[5] for the creation, visualization, and manipulation of our ontology "Parkinson's Disease Ontology." Protégé platform provides a growing user community with a suite of tools to construct domain models and knowledge-based applications with ontologies. Protégé can be customized to provide domain-friendly support for creating knowledge models and entering data.

---

[3] www.geneontology.org

[4] www.biopax.org

[5] http://protege.stanford.edu/download/protege/3.4/installanywhere/Web_Installers/

17.3 Results and Discussion

17.3.1 System Implementation in Protégé

The proposed Protégé knowledge base is consisted of the Protégé ontology "Parkinson's Disease Ontology" and the individual instances of the already mentioned classes with specific values for our roles. The graphical representation of the above classes is given in Fig. 17.4.

The proposed knowledge tree has six levels depth named as PD00, PD01, PD02, PD03, PD04, and PD05 respectively. PD00 represents the leaves of the tree. Each node of the represented knowledge tree denotes a Protégé class. For the sake of clarity, not all vertices were fully expanded. The relationships between classes are of type is-a (the child class is a type of the parent class) and they are represented by the blue lines in Fig. 17.5.

We define the properties of our classes using roles named VarPD _PD00, VarPD _PD01, VarPD _PD02, VarPD _PD03, VarPD _PD04, and VarPD _PD05. Each role corresponds to each class level in the knowledge tree. The value type of each role is Symbol and the cardinality is multiple. Also, we have specified the allowed values for the role. These values provide available symbols for individual instances of the mentioned classes; each of these symbols represents a propositional variable (Fig. 17.5). The red lines represent the instance of relation.

Examples of these instances are given below:
The role VarPD _PD00 includes the classes Bradykinesia, Tremor, Unsteady, Anxiety, Apathy, Binge Eating, Dental Disorders, Obsessive-Compulsive

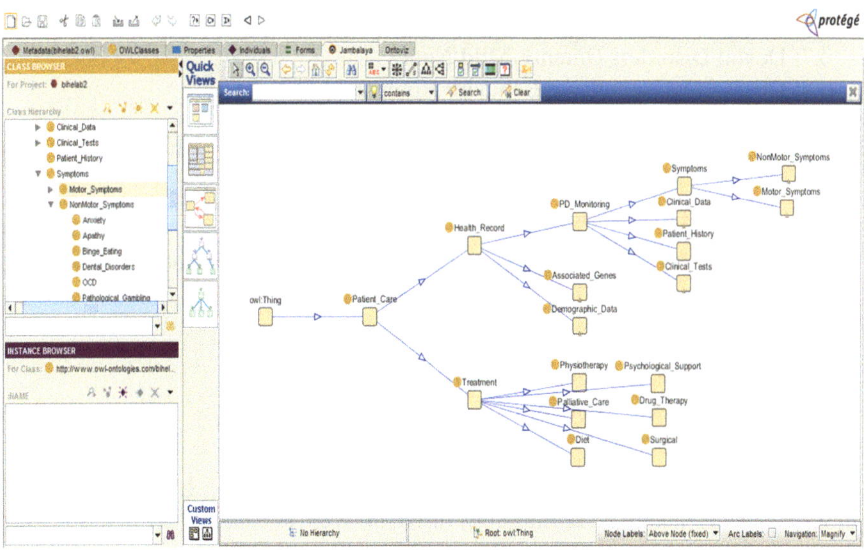

Fig. 17.4 The graphical representation of the classes in Protégé

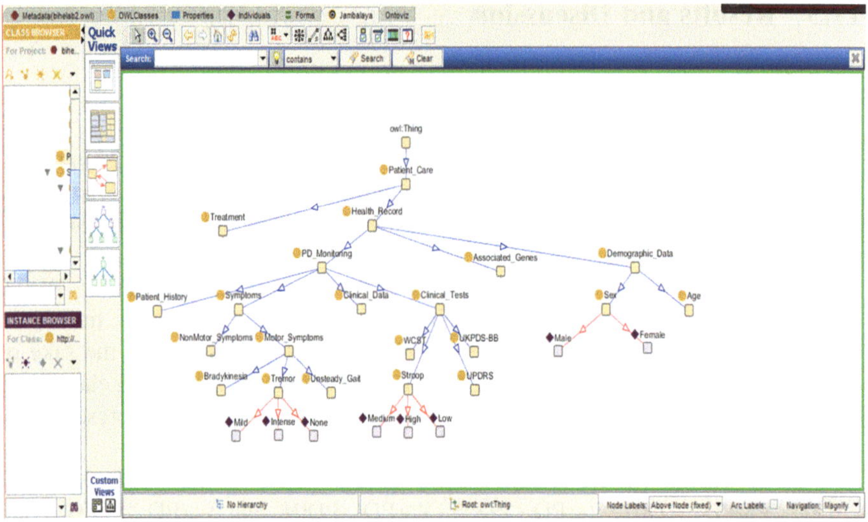

Fig. 17.5 The graphical representation of instances of classes Tremor, Stroop, and Sex in Protégé

Behavior, Pathological Gambling, Psychotic Symptoms, Vision Disorders, and Weight Loss or Gain. The instances of these classes are:

- x1: Bradykinesia (Intense/Mild/None)
- x2: Tremor (Intense/Mild/None)
- x3: Unsteady Gait (Intense/Mild/None)
- x4: Anxiety (Intense/Mild/None)
- x5: Apathy (Intense/Mild/None)
- x6: Binge Eating (Intense/Mild/None)
- x7: Dental Disorders (Intense/Mild/None)
- x8: Obsessive-Compulsive Behavior (Intense/Mild/None)
- x9: Pathological Gambling (Intense/Mild/None)
- x10: Psychotic Symptoms (Intense/Mild/None)
- x11: Vision Disorders (Intense/Mild/None)
- x12: Weight Loss or Gain (Intense/Mild/None)

The role VarPD _PD01 includes among others the classes Stroop, UKPDS-BB, UPDRS, and WCST. The instances of these classes are:

- x20: Stroop (high/medium/low)
- x21: UKPDS-BB (high/medium/low)
- x22: UPDRS (high/medium/low) (high/medium/low)
- x23: WCST (high/medium/low)

OWL[6] is a language based on XML supported by the W3C. It aims to formally describe the meaning of terminology used in Web documents so that software

[6] http://www.w3.org/TR/2004/REC-owl-features-20040210/#s1.2

```
<owl:Class rdf:about="#Treatment">
        <rdfs:subClassOf rdf:resource="#Patient_Care"/>
        <owl:disjointWith rdf:resource="#Health_Record"/>
</owl:Class>

<owl:Class rdf:about="#Physiotherapy">
        <owl:disjointWith rdf:resource="#Palliative_Care"/>
        <owl:disjointWith rdf:resource="#Diet"/>
        <owl:disjointWith rdf:resource="#Surgical"/>
        <owl:disjointWith rdf:resource="#Psychological_Support"/>
        <owl:disjointWith rdf:resource="#Drug_Therapy"/>
        <rdfs:subClassOf rdf:resource="#Treatment"/>
</owl:Class>

<owl:Class rdf:about="#Palliative_Care">
        <owl:disjointWith rdf:resource="#Physiotherapy"/>
        <owl:disjointWith rdf:resource="#Diet"/>
        <owl:disjointWith rdf:resource="#Surgical"/>
        <owl:disjointWith rdf:resource="#Psychological_Support"/>
        <owl:disjointWith rdf:resource="#Drug_Therapy"/>
        <rdfs:subClassOf rdf:resource="#Treatment"/>
</owl:Class>
```

Fig. 17.6 The resulting OWL file for the classes Treatment, Physiotherapy, and Palliative_Care

programs are able to easily process that information. Three main sublanguages have emerged from the standard OWL syntax: OWL-Lite, OWL-DL, and OWL-Full. Protégé provides an extension to its standard implementation that supports the Web Ontology Language, Protégé-OWL. Protégé-OWL[7] has been used to develop Parkinson's Disease Ontology and visualize the resulting class hierarchy (Fig. 17.6).

For the assessment of the patient's status with the appropriate treatment we design an expert system using a multilayer neural network. Multilayer networks solve the classification problem for nonlinear sets by employing hidden layers, whose neurons are not directly connected to the output. The additional hidden layers can be interpreted geometrically as additional hyperplanes, which enhance the separation capacity of the network.

The assumed architecture includes five feed-forward layers. The input layer corresponds to each lower-level subclass of the ontology tree, where each class' instances represent the domain of the respective input. Three hidden layers are connected to the input layer. Each hidden layer is linked with its neighboring neurons following the structure of inheritance in the knowledge tree (PD_Monitoring, Associated_Genes, Demographic_Data). Finally the output layer is fully connected to the previous one, corresponding to each available treatment.

For example, consider an input vector corresponding to a patient with intense tremor, mild bradykinesia, and mildly unsteady gait as well as a combination of other symptoms (Fig. 17.7). These data will be fed forward to the next layer for the

[7] http://protege.stanford.edu/download/download.html

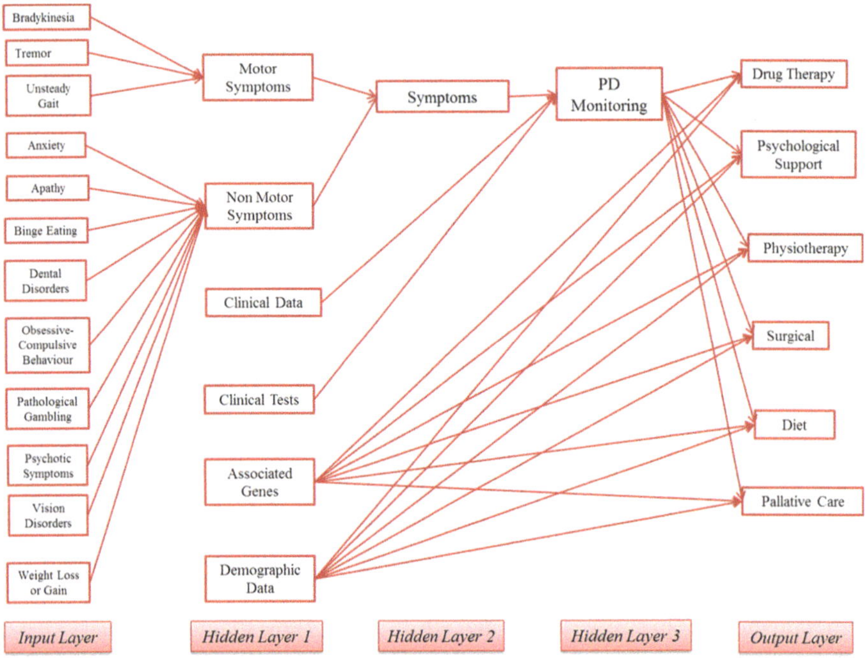

Fig. 17.7 An example of the proposed multilayer neural network. The inputs are PD symptoms and the outputs are the corresponding treatments

network to evaluate and assess the severity of the patient's motor and non-motor symptoms. This process will subsequently be followed to the second layer to assess the general severity of the symptoms and to the third layer where taking into consideration the patient's clinical tests, clinical data, and symptoms the overall state of the patient's monitoring will be estimated. From this point, data from other sources like the patient's associated genes and his demographic data will be combined to propose a treatment based on the patient's health state.

Conclusions

PD is recognized as one of the most common neurologic disorders, affecting approximately 1 % of individuals older than 60 years. It is obvious that future treatments are born of today's unmet needs. In this paper, the general architecture of a hybrid expert system for accurate diagnosis of PD and for progress monitoring is proposed. The tool is an integrated ontology-based model including information about numerous biomarkers, factors, and possible treatments. It will make use of patient data to provide patient-specific assessments or recommendations to clinicians to assist with clinical decision making. It will also aim to generate new knowledge of biomarkers and other

(continued)

(continued)

biofactors associated to PD. To achieve these goals, a combined effort incorporating information from genetic/clinical data, development of innovative integrated systems that combine multiparametric data, and application of complex algorithms is required.

The proposed systems will help physicians to decide more easily and faster what treatment to give to patients with PD, providing different treatment options in individual patients. In future work, we will use supervised learning techniques for data training in the proposed neural network and we will validate the proposed model using retrospective patient data.

Conflict of Interest The authors declare that there is no conflict of interest regarding the publication of this article. The authors of this manuscript do not have a direct financial relation that might lead to a conflict of interest for any of them.

References

1. Han M, Nagele E, DeMarshall C, Acharya N, Nagele R (2012) Diagnosis of Parkinson's disease based on disease-specific autoantibody profiles in human sera. PLoS One 7(2):e32383
2. Schapira AHV (2006) Etiology of Parkinson's disease. Neurology 66(4):S10–S23
3. Martin LJ (2010) Mitochondrial and cell death mechanisms in neurodegenerative diseases. Pharmaceuticals 3:839–915
4. Alexiou A, Rekkas J, Vlamos P (2011) Modeling the mitochondrial dysfunction in neurogenerative diseases due to high H+ concentration. Bioinformation 6(5):173–175, PubMed: 21738307
5. Alexiou A, Vlamos P (2012) Evidence for early identification of Alzheimer's disease, arXiv:1209.4223v2 [q-bio.NC]
6. Chan DC (2007) Mitochondrial dynamics in disease. New Engl J Med 356(17):1707–1709
7. Chen H, Chan DC (2009) Mitochondrial dynamics - fusion, fission, movement, and mitophagy - in neurodegenerative diseases. Hum Mol Genet 18(2):R169–R176
8. Howlett DR (2003) Protein misfolding in disease: cause or response? Curr Med Chem Immun Endoc Metab Agents 3:371–383
9. Hauser RA, Grosset DG (2012) [123I]FP-CIT (DaTscan) SPECT brain imaging in patients with suspected Parkinsonian syndromes. J Neuroimaging 22:225
10. Grosset D, Taurah L, Burn DJ, MacMahon D, Forbes A, Turner K (2007) A multicentre longitudinal observational study of changes in self-reported health status in people with Parkinson's disease left untreated at diagnosis. J Neurol Neurosurg Psychiatry 78(5):465–469
11. Grimes DA, Lang AE (1999) Treatment of early Parkinson's disease. Can J Neurol Sci 26 (2):39–44
12. Thobois S, Delamarre-Damier F, Derkinderen P (2005) Treatment of motor dysfunction in Parkinson's disease: an overview. Clin Neurol Neurosurg 107(4):269–281
13. Suchowersky O, Reich S, Perlmutter J, Zesiewicz T, Gronseth G, Weiner WJ (2006) Practice parameter: diagnosis and prognosis of new onset Parkinson disease (an evidence-based review), report of the Quality Standards Subcommittee of the American Academy of Neurology. Neurology 66(7):968–975

14. Weintraub D, Burn DJ (2011) Parkinson's disease: the quintessential neuropsychiatric disorder. Mov Disord 26(6):1022–1031
15. Braak H, Del Tredici K, Rüb U, de Vos R, Jansen SE, Braak E (2003) Staging of brain pathology related to sporadic Parkinson's disease. Neurobiol Aging 24:197–211
16. Verhagen LM (2008) Early diagnosis of PD: let the patient tell you, Johns Hopkins Advanced Studies in Medicine
17. Jankovic J (2008) Parkinson's disease: clinical features and diagnosis. J Neurol Neurosurg Psychiatry 79:368–376
18. Bayulkem K, Lopez G (2011) Clinical approach to nonmotor sensory fluctuations in Parkinson's disease. J Neurol Sci 310(1–2):82–855
19. Kleiber M, Kulpa Z (1995) Computer-assisted hybrid reasoning in simulation and analysis of physical systems. CAMES 2(3):165–186
20. Rubin DL, Talos I-F, Halle M, Musen MA, Kikinis R (2009) Computational neuroanatomy: ontology-based representation of neural components and connectivity. BMC Bioinform 10(2):S3

Chapter 18
Kinesia Paradoxa: A Challenging Parkinson's Phenomenon for Simulation

Eirini Banou

Abstract The present work aims to study extensively the literature on the phenomenon of "kinesia paradoxa" presented in Parkinson's disease patients, who generally cannot move but under certain circumstances exhibit a sudden, brief period of mobility (walking or even running). The objective of this study was to identify the mechanisms causing this phenomenon and relate them with respectively computational simulations aiming to draw attention to gaps and weaknesses and possible directions for future research. The study of this phenomenon with the use of modeling techniques may be a decisive factor for its interpretation.

Keywords Noradrenergic • Basal ganglia reserves • Cerebellar circuit • Visual cues • Parkinson's disease • PK • PD

18.1 Introduction

"Kinesia Paradoxa" or "Paradoxical Kinesia" or "Kinesia Paradoxica" or "Kinesie Paradoxale" (PK) is a term coined by Achille Alexandre Souques, one of the most famous French neurologists of the early twentieth century to describe "a sudden and brief period of mobility typically seen in response to emotional or physical stress" [1, 2] in patients with advanced idiopathic Parkinson's disease (PD). Another definition is given by Oliver Sacks [3]. Moreover, PK [4] was defined as "the sudden transient ability of a patient with Parkinson's disease (PD) to perform a task he or she was previously unable to perform."

The events recorded in the literature [5, 6] are really impressive: i.e., PD patients unable to walk suddenly run to save themselves from a fire.

On the other hand, some researchers support [7] that PK is not a Hallmark of PD but a general property of the motor system.

According to the literature (1) PK has been observed in other diseases like autism and Asperger's disease [8, 9] and (2) PK is also present to speech [10]: one patient with dysarthria—if encouraged—will repeat single words loud and clear.

E. Banou (✉)
Department of Informatics, Ionian University, 7 Tsirigoti Square, Corfu, Greece
e-mail: irene.banou@gmail.com

P. Vlamos, A. Alexiou (eds.), *GeNeDis 2014*, Advances in Experimental Medicine and Biology 822, DOI 10.1007/978-3-319-08927-0_18

18.2 Causes of Kinesia Paradoxa

If we would like to categorize PK based on its causes, three mechanisms are generally proposed: (a) life-threatening events (b) external stimuli, and (c) medication.

18.2.1 PK Induced by Life-Threatening Events

The most famous story [11] about this fascinating phenomenon is about a fire in a hospital in which the PD patients were trapped on the third floor. The nurses explained this to the firemen. Suddenly someone noticed that the wheelchair-bound PD patients were all outside the building. They escaped the fire by running downstairs.

Researchers reported [12] that during a strong earthquake in Italy in the city of L'Aquila one 90 years old woman, bedridden for a year before the earthquake, stood up from the bed and ran out of her apartment on the third floor. Fifteen days after the quake, she was walking assisted by her daughter, while in the previous visit to the clinic she had come on a wheelchair. Moreover, all 14 PD patients included in that study not only were able to escape from the buildings that collapsed (most of them build 100 years before) of the 6 Richter scale earthquake but helped and saved the lives of carers and/or members of their families.

Neurologist's doctors from Haifa, Israel [13] studied PK occurrence to PD patients in a war zone—at the northern part of Israel—where rockets were falling after sirens sounded. One of the patients experienced PK as a result of war circumstances. He has been suffering from Freezing Of Gait (FOG) for up to 10 years and he was unable to run. He did not react to the sound of siren. But after his wife grabbed him, he started to run following her footsteps.

A PD completely akinetic patient [14] was at home babysitting his 6-year-old grandson. The boy went out to play without his grandfather supervision. Suddenly a loud noise from a car accident was heard. The grandfather feared that his grandson has been run over by a car and immediately ran to the front door when he realized that the boy was safe but someone else had been hurt.

However, PK induced by life-threatening events cannot be studied experimentally (i.e., in real conditions) [15, 16] directly and systematically due to ethical constraints.

18.2.2 PK Induced by External Stimuli (Cueing)

18.2.2.1 Visual Cues

James Purdon Martin has conducted one of the first studies [17] on PK in 1967. He examined the effects of visual cues in PD patients walking ability. Specifically, patients were asked to walk along a runway with a variety of visual cues (i.e., blocks, tape, paper, etc.) placed upon it. The conclusions were that patients could sustain near normal walking in the presence of bold lines or an obstacle. Transverse lines were more effective than zig-zag lines and bold lines with contrasting color were also more effective.

Morris et al. [18] used visual cues (floor markers) placed on the pathway which was tiled in with grey-green linoleum and concluded that PD patients could achieve the same stride as in the controls when they used visual cues. In their next study the same researchers [19] used also attention strategies that have been also effective. Other scientists [20] tried to identify the kind of visual cues (static or dynamic) required for normal gait. They concluded that the visual perception of motion affects motor ability in PD patients rather than position or orientation. Many other studies [21] have demonstrated the benefits of visual stimuli, but the interpretation of this phenomenon is controversial.

Regarding visually induced PK both the importance and volume of relative literature are very extensive that in our opinion should become a special research object.

18.2.2.2 Music, Dancing, and Other Auditory Stimuli

Dr. Sacks in "Musicophilia" [22] discusses the effect of music therapy on PD patients, reporting very impressive effects. Music is very important for those who have mobility problems. Surveys [16, 23–25] indicate that rhythmic sounds and gauges improve the speed and stride length and minimize FOG in PD patients. As a simple explanation [26, 27] for the observed results is that the auditory stimuli act as pacemaker function and provide an external rate which is able to stabilize the defective rate within the basal ganglia. According to a research published in "Nature Neuroscience" [28] dopamine levels seemed to rise by 9 % when volunteers listened to music they liked. That research also showed that music is inextricably bound up with the deepest reward system.

If the music is enough to "unlock" PD patients, the combination of music and movement that dance offers is undoubtedly beneficial. Hackney M. and G. Earhart, from the Washington School of Medicine, have documented in their studies [29, 30] the direct effects of dance: improved functional mobility and self-confidence. Researchers believe that the benefits of dance are extensive enough to be clinically meaningful. There are specialized dance programs [31] for patients with PD through which professional dancers try to improve patient's movement.

18.2.2.3 Cycling

Dr. Bastiaan R. Bloem of the Radboud University Nijmegen Medical Center in the Netherlands writes about a 58-year-old PD patient who was presented to him with severe FOG [5] unable to walk. The total surprise was that he had preserved the ability to ride a bicycle. He rode for several miles every day and while riding his bike all his Parkinson's symptoms were gone. After he dismounted the bicycle, he returned to his previous state (FOG). Generally, Parkinson's experts were excited about this observation and besides the fact that is a single-case phenomenon they believe that can provide crucial evidence.

18.2.3 Medication (L-Dopa)

Dr. Oliver Sacks, professor of clinical neurology, was a doctor at the Hospital of chronic illness of Mount Carmel, New York. There he met for the first time PD patients suffering from a pandemic—called Encephalitis Lethargica—who were imprisoned there after the First World War. For a long time these patients were in a strange situation of lethargy and they were considered almost dead. After treatment with drug L-DOPA, they began to metamorphose, to "resurrect" by resuming speaking and walking. They regained the joy of life and "woke up" from their lethargy. Sacks obsessively recorded these patients' behavior and asked them to keep notes about their state themselves. His book "Awakenings" [3] includes an anthology of these descriptions as well as interesting approaches. Some researchers claim [32] that the PK phenomenon occurs only under the influence of treatment with L-Dopa.

18.3 Related Work: PK Hypotheses (Approaches to Mechanisms)

Various theories have been developed to explain the PK phenomenon, focused on three fundamental axes: Activation of basal ganglia reserves, Activation of alternative pathways via Cerebellar Circuit, or via Noradrenergic Augmentation.

18.3.1 The Noradrenergic System

The Noradrenergic System organizes the body's response to stress in cooperation with other body systems in order to ensure survival [33]. Under a life-threatening event (e.g., fire or earthquake) even PD patients who couldn't walk managed to run out and this may be due to noradrenergic activation [34]. Many theories [35] argue that a deficiency in the NA system of the brain plays a crucial role in the progression of neurodegenerative diseases like PD and Alzheimer. NA constitutes the neural sub-layer of reaction readiness for fear-anxiety. Its role is to increase attention [36]. It is believed to be involved in the pathophysiology of pain and depression. Abnormalities in NA system lead to disorganization. The role of oxidative stress for human neurodegenerative diseases was also highlighted by the authors in [37, 38].

Other such systems are Dopaminergic system and Serotoninergic system. The major neurotransmitter localized in noradrenergic neurons is norepinephrine (NE), also known as noradrenaline (NA) [35]. NE or NA is a catecholamine. Catecholamines are also referred to as "fight or flight" hormones because they are released in response to stress. NA has been associated with several brain functions, including learning and emotions. According to Cannon [39] the brain responds to all emergencies in the same manner: increasing the adrenaline levels. Human body mobilizes all its energy in cases of severe anxiety or a threat stimulating the sympathetic nervous system, causing a kind of "general alarm."

18.3.2 Experiments and Studies

Noradrenergic augmentation as part of "fight or flight" response to stress or otherwise "adrenal medulla activation by environmental stress" [34] is one of the mechanism proposed to explain PK. This opinion was supported by several studies [11–13] and even by laboratory experiments with rats and with a monkey.

J. Jankovic [36] supports that bradykinesia is dependent on the emotional state of the PD patient. An unexpected change of emotional energy can transform the immobile patient such as to catch a ball or do other fast movements. This demonstrates that the motor programs are intact but the PD patients cannot use them without an external arousal.

In a study conducted in 1987 [34], it was investigated whether the adrenal activation from environmental stress affects the expression of catalepsy caused by neuroleptic drugs in rats. The stress situation was caused by forced immobilization (gauze bandage wrapping), exposure to cold (30 °C), and handling (rats were placed in hand and continuously moved for 20 min). Catalepsy was induced by Haloperidol. The results showed that these Stress Procedures reduced the catalepsy conditions. Some rats did not react and this may be explained because they have different states in their adrenal gland. Researchers link their findings with PK and support them by mentioning other studies where the lack of adrenal medulla activity contributes to clinical symptoms of idiopathic or drug-induced parkinsonism. They conclude by suggesting that during long-term neuroleptic medication, behavioral tolerance is independent of brain dopaminergic processes but depends on memory factors and emotional stress.

The theory of NA involvement in PK is also supported by C. Colpaert [40] who observed PK in reserpine treated rats when he exposed them to stress by throwing them in a water basin.

Moreover, in another study, Colpaert et al. [41] discuss the case of a 20-year-old java monkey who was treated with MPTP, a neurotoxin which, as observed in drug users, causes a syndrome resembling Parkinson's (rigidity, tremor, postural instability, hypokinesia, bradykinesia). Then both behavior and symptoms that the monkey exhibited were studied. In general, 7 months' observations showed that MPTP induced tremor and many others symptoms of PD to the animal. Among others, when faced with fear, monkey exhibited PK: On day 111, Jumbo—the monkey—was rigid and mute as usual, till he saw Helga, a female German Shepherd. Suddenly Jumbo got up, with perfect motor control (instability and mute disappeared) and started to run, growling loudly in order to attack Helga. The examiner stopped him and Jumbo retreated to his previous postural situation. Dogs and monkeys are natural enemies (prey), so this scene could be a PK phenomenon induced by stress.

18.3.3 The Basal Ganglia Circuit

Although the function of the basal ganglia is complex and largely unknown, we know that normal operation of basal ganglia circuit is required in order to have normal motion ability. But: "The basal ganglia have all the clarity of a dark basement" [42]. PD is caused by a degeneration of dopaminergic neurons in the substantia nigra pars compacta (SN_C). The substantia nigra [43] is one of the most significant four principal nuclei of the basal ganglia among the others: (1) the striatum, (2) the globus pallidus, and (3) the subthalamic nucleus. The striatum consists of three important subdivisions: the caudate nucleus, the putamen, and the ventral striatum (which includes the nucleus accumbens). Thalamus, cerebral cortex, and basal ganglia (or basal nuclei) play a major role in normal voluntary movement [44]: "Basal ganglia have distinct pathways that compete with each other to release movement or to inhibit movement. The competing pathways act like the brake and accelerator in a car. The brake–accelerator model suggests that release (disinhibition) of the thalamus by the direct pathway is opposed by the indirect pathway, which inhibits the thalamus via the additional, excitatory, subthalamic projection to the internal pallidum."

Nambu et al. [45] added one more pathway by introducing the term "cortico-subthalamo-pallidal hyperdirect pathway" as a fast route linked with motor programs.

Dopamine [46] is produced by the cells of the pars compacta of the substantia nigra whose axons project into the striatum. The effect of dopamine is different in the direct and indirect pathway: Dopamine in the direct pathway enhances movement while in the indirect pathway it inhibits it. Accordingly, the action of dopamine in the direct and indirect pathway is regulatory for movement. Dopamine deficiency in the striatum—as occurs in PD—results in hyperactivity of the output area, and increased inhibitory activity in the thalamus thus reducing mobility.

18.3.4 Experiments and Studies

Several studies have shown that PD patients even in advanced stages can increase the release of dopamine. De la Fuente-Fernández et al., assuming that the placebo may have a prominent place in the treatment of Parkinson, wondered if the use of the placebo effect produced by activation of the pathway that has been damaged by degeneration (nigrostriatal dopaminergic system) and tried to answer this question using PET (Positron Emission Tomography). The magnitude of the response to placebo was comparable to that of the therapeutic dose of apomorphine or levodopa. Thus they found that the release of dopamine in nigrostriatal system is associated with the expectation of reward, and in this case in anticipation of therapeutic benefit. They concluded that the level of expectations can determine the experience: patients who felt effect from placebo had a higher release of dopamine from others. In later studies, same researchers [47] show that the expectation of reward—not the reward itself—triggers the release of dopamine in the nucleus accumbens. Moreover, because dopamine is the main neurotransmitter for reward circuit, the model predicts that the release of dopamine in the ventral striatum may be involved in the response to placebo, not only in parkinsonism but also in other medical conditions. Later, they suggested [48] a model of the placebo effect where the main element is dopamine. The model creates a link between the placebo and reward mechanisms. It is remarkable that patients suffering from Alzheimer's do not have the capacity to be affected by the placebo [49] because of lost dependent prefrontal cortex ability to have expectations.

Dopamine release in the striatum in response to placebo was confirmed [50] using rTMS (transcranial magnetic stimulation) in PD patients of advanced stage and also emphasized the correlation between reward mechanisms and PD.

18.3.5 Cerebellar Circuit (Alternative Pathways)

The cerebellum is very important in the preparation or execution of movement and is also involved in the correct timing of movements which is important for self-regulated movements.

Wu et al. argue [51] that the cerebellum has two roles in Parkinson's disease: pathophysiological (anatomical and other clinical evidence, i.e., tremor) and compensatory. The compensatory role defined by Galley [52] as "any morphological or functional change in the brain that is damaged—which change acts to maintain the performance of impaired functioning."

PD patients (OFF medication) show hyperactivity in the cerebellum, not only during the execution of motor activities (e.g., upper limb movements) but also at rest. The nature of hyperactivity and increased connectivity in the cerebellum in PD patients remains unclear. Many researchers support that this is a compensatory benefit [7].

According to studies [53–55] the cerebellum and basal ganglia have a two-way communication between them and are connected together to form an integrated functional network. The discovery of this mutual association between basal ganglia and cerebellum provides an anatomical basis for the interpretation of the role of the cerebellum in PD.

It is argued [4] that a specific visual–motor pathway can bypass the damaged basal ganglia and allow the intact cerebellar circuit used for visual motor control. This view of the role of the cerebellum as an alternative motor pathway was confirmed from Marsden and Obeso [56] and Rascol et al. [57].

18.4 Simulations (Literature Review)

18.4.1 Simulation Models of the Noradrenergic System

Because serious ethical barriers are placed for the experimental study of patient's response in life-threatening situations—in order to export conclusions about the role of the noradrenergic system in PK—the simulation in this case would constitute a very interesting challenge. The existing literature on this particular area (simulated responses to stressful situations, "fight or flight" response) does not have any references about the desired—for this work—role of the noradrenergic system. Overall, however, the neurotransmitter mechanisms have been studied through experimental simulations using artificial intelligence methods (machine learning, reinforcement learning) and Robotics for factors such as human cognition, learning, and rewarding experience. One simulation could be argued that approximates the axis of this work [58] and is about a model in which the agent has to respond appropriately to expected and not expected uncertainty. The response of the cholinergic (expected uncertainty) and the noradrenergic system (unexpected uncertainty) was recorded and a possible explanation for the relationship between both the locus coeruleus (LC) and basal forebrain with environment of uncertainty was given.

Yu and Dayan [59] built a similar model, based on the Bayesian Statistical model focusing on the role of the cholinergic and noradrenergic system on attentional tasks.

18.4.2 Basal Ganglia models

The fact that basal ganglia have a critical role in movement disorders—including PD—several computational models of basal ganglia were constructed. The qualitative approach (internal connections between basal ganglia and external brain structures) [60] and the quantitative approach: (a) low level that is cellular- and network-level models and (b) high-level models. Most of them have been developed for the main type of surgery used in Parkinson (Deep Brain Stimulation-DBS).

18.4.3 Simulation Models of Cerebellum

According to [61], we have two categories of cerebellar simulations (a) Top-down and (b) Bottom-up. In top-down approach, we begin with computational hypotheses and then try to find how these mechanisms might operate within the cerebellum. In bottom-up approach, we take known facts (synaptic, cellular) in order to build a simulation as accurate as possible. The cerebellum does one thing: accepts inputs through clearly defined afferent fibers and produces results in accordance with the internal rules of information processing. So, if we focus on what the cerebellum calculates we can greatly facilitate our understanding of how different cerebellar regions contribute to various cognitive and motor functions.

Mainly, computer simulations of eyelid conditioning try to explain (1) the relation between the motor learning and cerebellum and (2) contribute to cerebellum input–output transformation as mentioned above.

Computational models of other studies [62] focused on precise timing trying to investigate whether the plasticity in the cerebellar cortex alone can mediate on timing response. The same researchers in their next study [63] provide evidence supporting the view that cerebellar learning generally involves plasticity in the cerebellar cortex.

Yamazaki et al. [64] made a model from another point of view: The cerebellum as a liquid state machine.

SENSOPAC [65, 66] stands for "SENSOrimotor structuring of Perception and Action for emergent Cognition" and is an EU project, which combines machine learning techniques and knowledge about biological perception-action mechanisms. Within this project an artificial arm and hand were built. In order to provide this arm with intelligence, scientists decided to develop a cerebellum-based control system.

HUMAN BRAIN PROJECT [67] funded by the EU aims to simulate the complete human brain which seems to be nowadays a big challenge for scientists. Six research platforms (brain simulation, neuroinformatics, medical informatics, high-performance computing, neuromorphic computing, and neurorobotics) are promoted for setting up and testing. In addition, we must mention the pessimistic view formulated by G. Hinton [68] that "brain simulation will fail, because it consists too many moving parts that no one yet understands."

Conclusion: Future Work

The view of the noradrenergic system simulation in threatening or stressful situations—that would be extremely interesting for our subject—has not been extensively investigated. If we take into consideration that there are obvious ethical limitations in studying life-threatening situations in real conditions, it is immediately concluded that the computational modeling and simulation (noradrenergic reaction under stress and/or threat) is an ideal challenge for future research. In combination with others innovative ontology-based modeling tools [69] maybe we can develop new challenging patient-specific PD assessments.

In addition, the terms visual motor alternative pathway gait provided through the cerebellum are particularly interesting for further research, so far as a tendency for researchers to "gravitate" towards this view appears. The connection between PD gait and cerebellum certainly needs further investigation and extensively recording, organizing, and managing relative information. Moreover, the existing computational model should be further investigated in greater depth in order to design a model adapted to the specific requirements always regarding Paradoxical Kinesia.

One of the main questions that arise is why not all patients have the same response to various stimuli or situations? Apart from the different state of the adrenal gland mentioned above, possible answers could be: (a) that everybody responds differently to stress, if we accept the theory of increased noradrenergic function and (b) brain plasticity, i.e., the ability of brain to change its structural and functional organization in such a way that it can adapt to new functional requirements [70]. A scientifically based theory of the brain is the theory of redundancy. According to this, the brain is able to replicate neural pathways during development. Thus when one pathway is damaged, there is one that can take on the task.

These assumptions are leading to other scientific directions for future research in order to provide new evidence that will enhance any theory views or weaken others.

References

1. Souques AA (1921) Kinesie paradoxicale. Rev Neurol 37:559–560
2. Souques AA (1921) Rapport sur les syndromes parkinsoniens. Rev Neurol 28:534–573
3. Sacks O (1991) Awakenings. Pan Macmillan, London
4. Glickstein M, Stein J (1991) Paradoxical movement in Parkinson's disease. Trends Neurosci 14(11):480–482
5. Snijders AH, Bloem BR (2010) Cycling for freezing of gait. N Engl J Med 362(13):e46

6. The Michael J. Fox Foundation for Parkinson's Research, https://www.michaeljfox.org/foundation/publication-detail.html?id=86
7. Ballanger B et al (2006) "Paradoxical Kinesis" is not a Hallmark of Parkinson's disease but a general property of the motor system. Mov Disord 21(9):1490–1495
8. Rinehart N, McGinley J (2010) Is motor dysfunction core to autism spectrum disorder? Dev Med Child Neurol 52(8):697
9. Rinehart NJ et al (2006) An examination of movement kinematics in young people with high-functioning autism and Asperger's disorder: further evidence for a motor planning deficit. J Autism Dev Disord 36(6):757–767
10. Critchley EM (1981) Speech disorders of Parkinsonism: a review. J Neurol Neurosurg Psychiatry 44(9):751–758
11. Hammond TC (2010) New developments: falls, drooling & exercise in Parkinson's Disease. The Parkinson's Source (40). APDA magazine
12. Bonanni L et al (2010) Protracted benefit from paradoxical kinesia in typical and atypical parkinsonisms. Neurol Sci 31(6):751–756
13. Schlesinger I, Erikh I, Yarnitsky D (2007) Paradoxical kinesia at war. Mov Disord 22 (16):2394–2397
14. Daroff RB (2008) Paradoxical kinesia. Mov Disord 23(8):1193
15. Robottom BJ, Weiner WJ (2009) Kick and rush: paradoxical kinesia in Parkinson disease. Neurology 73(4):328–329
16. Anzak A et al (2011) Improvements in rate of development and magnitude of force with intense auditory stimuli in patients with Parkinson's disease. Eur J Neurosci 34(1):124–132
17. Martin JP (1967) The basal ganglia and posture. J.B. Lippincott Company, Philadelphia, PA
18. Morris ME et al (1994) Ability to modulate walking cadence remains intact in Parkinson's disease. J Neurol Neurosurg Psychiatry 57(12):1532–1534
19. Morris ME et al (1996) Stride length regulation in Parkinson's disease normalization strategies and underlying mechanisms. Brain 119(2):551–568
20. Azulay J-P et al (1999) Visual control of locomotion in Parkinson's disease. Brain 122(1):111–120
21. Kelly VE et al (2002) Interaction of levodopa and cues on voluntary reaching in Parkinson's disease. Mov Disord 17(1):38–44
22. Sacks O (2010) Musicophilia: tales of music and the brain. Chapter 20: Kinetic melody: Parkinson's disease and music therapy. Random House Digital Inc., New York, NY
23. Styns F et al (2007) Walking on music. Hum Mov Sci 26(5):769–785
24. Vella-Burrows T, Hancox G (2012) Singing and people with Parkinson's. Sidney De Haan Research Centre for Arts and Health, Canterbury Christ Church University, Canterbury
25. Arias P, Cudeiro J (2010) Effect of rhythmic auditory stimulation on gait in Parkinsonian patients with and without freezing of gait. PLoS One 5(3):e9675
26. Hausdorff JM et al (2007) Rhythmic auditory stimulation modulates gait variability in Parkinson's disease. Eur J Neurosci 26(8):2369–2375
27. Fernandez del Olmo M, Cudeiro J (2003) A simple procedure using auditory stimuli to improve movement in Parkinson's disease: a pilot study. Neurol Clin Neurophysiol 25:2003–2022
28. Salimpoor VN et al (2011) Anatomically distinct dopamine release during anticipation and experience of peak emotion to music. Nat Neurosci 14(2):257–262
29. Earhart GM (2009) Dance as therapy for individuals with Parkinson disease. Eur J Phys Rehabil Med 45(2):231–238
30. Hackney ME, Earhart GM (2010) Effects of dance on balance and gait in severe Parkinson disease: a case study. Disabil Rehabil 32(8):679–684
31. Houston S, McGill A (2013) A mixed-methods study into ballet for people living with Parkinson's. Arts Health 5(2):103–119
32. Hardie RJ (1990) Parkinson's disease, Chapter 20. Chapman and Hall Medical, London, pp 559–596

33. Marien MR, Colpaert FC, Rosenquist AC (2004) Noradrenergic mechanisms in neurodegenerative diseases: a theory. Brain Res Rev 45(1):38–78
34. Yntema OP, Korf J (1987) Transient suppression by stress of haloperidol induced catalepsy by the activation of the adrenal medulla. Psychopharmacology (Berl) 91:131–134
35. Szot P, Franklin A, Raskind MA (2011) The noradrenergic system is a major component in Parkinson's disease. Etiology and pathophysiology of Parkinson's disease. InTech Open Access, Rijeka, Croatia, pp 247–272
36. Jankovic J (2003) Pathophysiology and clinical assessment of Parkinsonian symptoms and signs. Neurol Dis Ther 59:71–108
37. Athanasios A, Rekkas J, Vlamos P (2011) Modeling the mitochondrial dysfunction in neurogenerative diseases due to high H+concentration. Bioinformation 6(5):173, PubMed: 21738307
38. Alexiou AT, Vlamos PM, Volikas KG (2010) A theoretical artificial approach on reducing mitochondrial abnormalities in Alzheimer's disease. Proceedings of the 10th International Conference on Information technology and applications in biomedicine: emerging technologies for patient specific healthcare (ITAB'10), Corfu, Greece, November 2010
39. Cannon WB (1932) The wisdom of the body
40. Colpaert FC (1987) Pharmacological characteristics of tremor, rigidity and hypokinesia induced by reserpine in rat. Neuropharmacology 26(9):1431–1440
41. Degryse A-D, Colpaert FC (1986) Symptoms and behavioral features induced by 1-methyl-4-phenyl-1, 2, 3, 6-tetrahydropyridine (MPTP) in an old java monkey [*Macaca cynamolgus fascicularis* (Raffles)]. Brain Res Bull 16(5):561–571
42. Wilson SK (1925) Croonian lectures. Lancet 2(1):53
43. DeLong MR (2000) The basal ganglia. Principles of neural science, vol 4. McGraw-Hill, New York, NY, pp 647–659
44. Graybiel AM (2000) The basal ganglia. Curr Biol 10(14):R509–R511
45. Nambu A, Tokuno H, Takada M (2002) Functional significance of the cortico–subthalamo–pallidal 'hyperdirect' pathway. Neurosci Res 43(2):111–117
46. Michmizos KP (2011) Development of computational and mathematical models of biological neurons for the study and the control of the pathophysiology of motion. Dissertation. NTUA
47. De la Fuente-Fernández R et al (2002) Dopamine release in human ventral striatum and expectation of reward. Behav Brain Res 136(2):359–363
48. De la Fuente-Fernandez R, Lidstone S, Stoessl AJ (2006) Placebo effect and dopamine release. Parkinson's disease and Related Disorders. Springer, Vienna, pp 415–418
49. Zubieta J-K et al (2006) Belief or need? Accounting for individual variations in the neurochemistry of the placebo effect. Brain Behav Immun 20(1):15–26
50. Strafella AP, Ko JH, Monchi O (2006) Therapeutic application of transcranial magnetic stimulation in Parkinson's disease: the contribution of expectation. Neuroimage 31(4):1666–1672
51. Wu T, Hallett M (2013) The cerebellum in Parkinson's disease. Brain 136(3):696–709
52. Galley SL (2012) A joint compensatory and default mode network closely related to motor performance in Parkinson's disease. Thesis
53. Doya K (2000) Complementary roles of basal ganglia and cerebellum in learning and motor *control*. Curr Opin Neurobiol 10(6):732–739
54. Doya K (1999) What are the computations of the cerebellum, the basal ganglia and the cerebral cortex? Neural Netw 12(7):961–974
55. Bostan AC, Dum RP, Strick PL (2010) The basal ganglia communicate with the cerebellum. Proc Natl Acad Sci 107(18):8452–8456
56. Marsden CD, Obeso JA (1994) The functions of the basal ganglia and the paradox of stereotaxic surgery in Parkinson's disease. Brain, 117(4):877–897
57. Rascol, O., et al. (1997) The ipsilateral cerebellar hemisphere is overactive during hand movements in akinetic parkinsonian patients. Brain 120(1):103–110

58. Avery MC et al (2012) Simulation of cholinergic and noradrenergic modulation of behavior in uncertain environments. Front Comput Neurosci 6:1–16
59. Yu AJ, Dayan P (2005) Uncertainty, neuromodulation, and attention. Neuron, 46(4):681–692
60. Nikita KS, Tsirogiannis GL (2007) Computational models simulating electrophysiological activity in the basal ganglia. Operative neuromodulation. Springer, Vienna, pp 505–511
61. Medina JF, Mauk MD (2000) Computer simulation of cerebellar information processing. Nat Neurosci 3:1205–1211
62. HofstoÈtter C, Mintz M, Verschure PFMJ (2002) The cerebellum in action: a simulation and robotics study. Eur J Neurosci 16:1361
63. Ohyama T et al (2006) Learning-induced plasticity in deep cerebellar nucleus. J Neurosci 26(49):12656–12663
64. Yamazaki T, Tanaka S (2007) The cerebellum as a liquid state machine. Neural Netw 20(3):290–297
65. Carrillo RR et al (2008) A real-time spiking cerebellum model for learning robot control. Biosystems 94(1):18–27
66. The Sensopac Project, http://www.sensopac.org/
67. Human Brain Project, https://www.humanbrainproject.eu/el
68. Natalie W (2013) As machines get smarter, Evidence they learn like us. Quanta magazine. Simons Foundation, New York, NY
69. Alexiou A, Psiha M, Vlamos P (2012) An integrated ontology-based model for the early diagnosis of Parkinson's disease. 8th Artificial Intelligence Applications & Innovations, H. Papadopoulos et al. (Eds.): AIAI 2012, IFIP AICT 382, pp. 442–450, 2012, © IFIP International Federation for Information Processing 2012, Springer, Heidelberg, Chalkidiki Greece
70. Bach-y-Rita, P (1980) Brain plasticity as a basis of the development of rehabilitation procedures for hemiplegia. Scand J Rehabil Med 13(2–3), 73–83

Chapter 19
Towards the Development of a Device to Improve the Gait of Patients Suffering from Parkinson's Disease

Nikos Tsotsolas

Keywords Assistive technology • Visual cues • Parkinson's disease • Kinesia paradoxa

Our team is developing a technologically advanced and low-cost device to support the people with Parkinson's disease in successfully overcoming the phenomenon of freezing (kinesia paradoxa) without external aid. The device, which is using embedded motion sensors, namely gyroscopes, accelerometers, and magnetometers, is connected to smart transparent glasses in order to display visual cues. It also includes an embedded software system that fully supports the integration and analysis of the several input signals supporting the generation of visual and audio stimulation which stimulate patients with Parkinson's disease to walk in their off-phase. The device meets all the requirements for improving the quality of life of people with Parkinson's disease.

N. Tsotsolas (✉)
P.A.N. ANTISTIXIS SA, Athens, Greece
e-mail: ntsotsolas@pan-antistixis.com

P. Vlamos, A. Alexiou (eds.), *GeNeDis 2014*, Advances in Experimental Medicine and Biology 822, DOI 10.1007/978-3-319-08927-0_19

Chapter 20
Adaptations in Aquatic Organisms for Increased Sensitivity to Light and Differences from Humans

Vasileios K. Petropoulos, Ioannis K. Petropoulos, and Georgios Verriopoulos

Abstract The retina of aquatic organisms has been adapted, through evolution, to the specific lighting conditions of water. The purpose of this paper is to present the major morphological and functional differences of photoreceptors between humans and aquatic organisms. Comparison of visual pathways of the nervous system between humans and aquatic organisms is also attempted.

Keywords Aquatic organisms • Light • Retina • Photoreceptors • Cones • Rods • Camouflage

20.1 Introduction

Visual impressions of aquatic organisms depend on the depth in which they live. There are two main natural sources of light in the sea: that provided by the sun, moon, and stars, and that of bioluminescence produced by aquatic animals. In the mesopelagic zone (depth 150–1,000 m), the sunlight becomes dimmer and bluer as the depth grows. In the bathypelagic zone (depth >1,000 m), the light is visible only through bioluminescence, i.e., through reflection or transmission by other organisms. Finally, at the bottom of the sea, the world is primarily two dimensional and food is a matter of rich organic deposit [1].

The retina of aquatic organisms has been adapted, through evolution, to specific lighting conditions in water, especially in relation to the particularities and requirements of depth in which they live [1].

V.K. Petropoulos (✉)
Ionian University, Corfu, Greece
e-mail: vas_pet@biol.uoa.gr

I.K. Petropoulos
Ophthalmological Center of Rive, Geneva, Switzerland

G. Verriopoulos
University of Athens, Panepistimioupoli, Zografou, Greece

P. Vlamos, A. Alexiou (eds.), *GeNeDis 2014*, Advances in Experimental Medicine and Biology 822, DOI 10.1007/978-3-319-08927-0_20

The purpose of this work is the presentation of the main morphological and functional differences of photoreceptors between the eye of man and that of aquatic organisms. Comparison of visual pathways of the nervous system between humans and aquatic organisms is also attempted.

20.2 Material and Methods

We present histological observations in the retinas of various species of aquatic organisms and attempted comparisons with humans.

20.3 Results

Both in humans and in aquatic organisms, cones are responsible for vision in bright light (photopic) and for the perception of color, whereas rods serve the need for vision in dim light (scotopic) [2] (Figs. 20.1 and 20.2). Yet most of the aquatic organisms have wider range of color perception than humans, while many fish are capable of perceiving ultraviolet light.

Additionally, as the light fades when traveling into the water, especially at great depths, most fish have developed sets of double cone cells or multiple layers of rods (Figs. 20.3 and 20.4), with a view to reinforce the perception of dim radiations [1]. Another evolutionary adaptation of fish is the adjustment of a photopigment absorption maximum to match the emission maximum of light radiation in the water where the fish lives. Therefore, at great depths, where blue rays penetrate more, fish photopigment has maximum absorption in the blue [3]. Finally, as the sunlight enters the water polarized, fish have 4 types of cone cells, in order to perceive the polarized light, to enhance it, and in some cases to use it as camouflage. Specifically, Hawryshyn and McFarland have demonstrated that ultraviolet-sensitive cones are activated by the vertically polarized light, whereas cones that are sensitive to green and red are activated by the horizontally polarized light. There are also organisms that, either by depolarizing or by changing the polarized field of light that passes through them, manage to change their visual characteristics for camouflage purposes [4] (Fig. 20.5).

The contrast sensitivity function of a subject shifts upwards to higher sensitivities after practice using spatial gratings [5]. Moreover, the ability to resolve low-contrast images is influenced by practice [6]. Since the human brain and visual system are highly plastic [7], enhanced underwater visual acuity of some children might thus be explained not only by learning to accommodate but also by changes in the visual pathways of the nervous system [8].

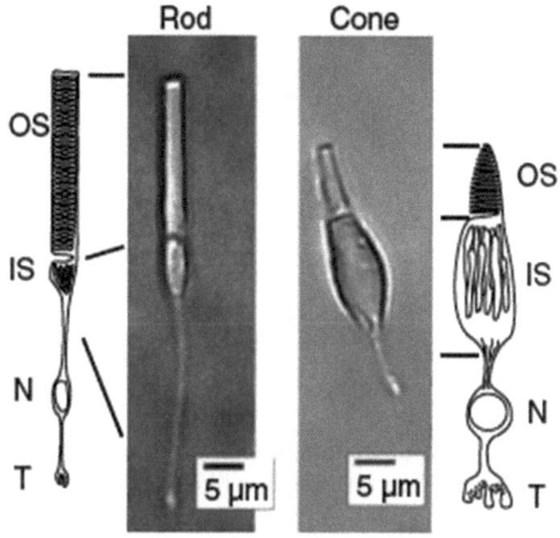

Fig. 20.1 Schematic representation of a rod (*left*) and a cone (*right*). Their relative size can be seen. From Tachibanaki et al. [9]

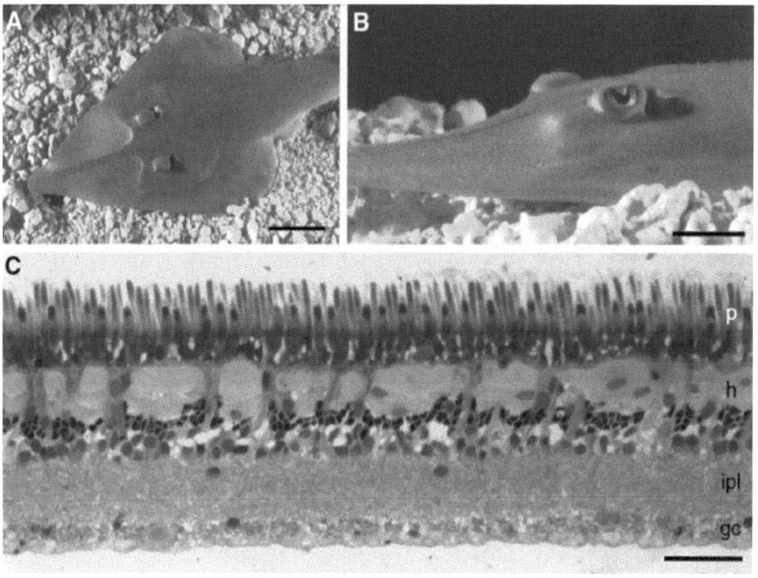

Fig. 20.2 Histological picture of photoreceptors in the retina layout of Rhinobatos typus elasmobranch. Alternation of cone and rod photoreceptors is observed. Cones are smaller in length. From Hart et al. [10]

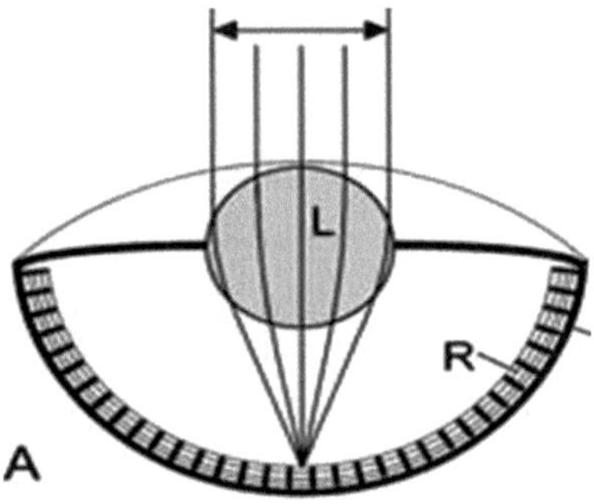

Fig. 20.3 All marine vertebrates and several marine invertebrates (Gastropods, Cephalopods, some Annelids, Crustaceans like Copepods) have camera-type eyes, namely eyes where the image is formed on a retina (R) passing through a lens (L). This allows for multiple magnifications of the light stimulus from the rods, adjustment that proves necessary for organisms that live in the bathypelagic zone (>1,000 m depth), where in addition to the particularly low lighting conditions, organisms are obliged to deal with two types of traffic noise: the "transducer's noise," namely electrical responses to increased biophosphorism, and the "dark noise," i.e., random light electrical responses from other biochemical processes. Visual "noise" reduces visual accuracy. Thus, multiple layers of rods of bathypelagic zone organisms constitute a valuable retinal adaptation to compensate for low lighting in that zone. From Nilsson [11]

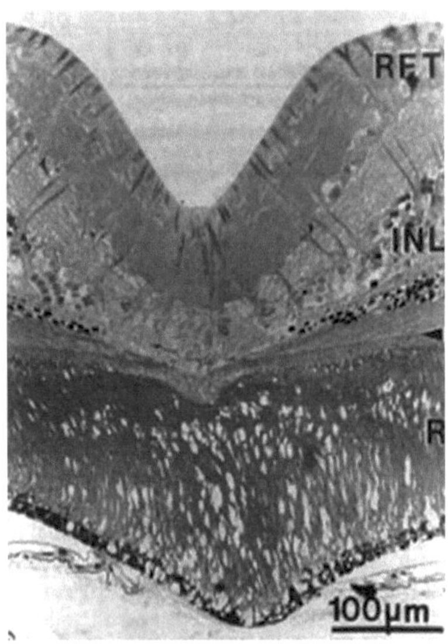

Fig. 20.4 One of the adaptations that bathypelagic organisms have developed refers to the number and arrangement of rods, exclusive photoreceptors of these organisms. The bathypelagic alepocephalid teleost Bajacalifornia drakei features 28–39 rod layers in each temporal fovea. The rods are only 27 µm each, but the multiple layers ensure visual pathway of approximately 750 µm. This extremely rare example occurs in other bathypelagic organisms as well, but with fewer rod layers. In this way organisms manage to enlarge the luminous impulse and to distinguish it from the aforementioned noises. From Wagner et al. [12]

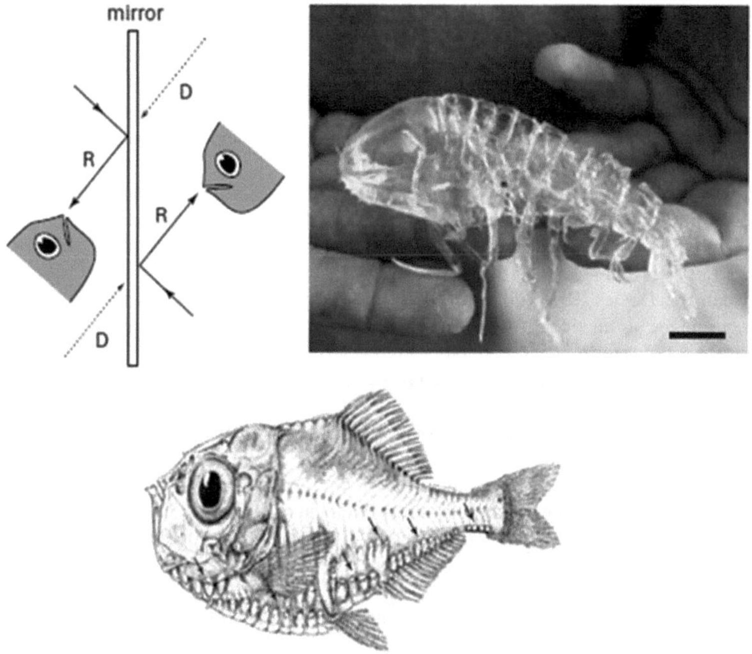

Fig. 20.5 Examples of camouflage techniques of fish. *Above:* mirror technique. A fish looking at a flat mirror cannot distinguish between the reflecting (R) and incidence (D) light beam, so the mirror becomes invisible. In water, that mirror may be another fish which, thanks to its smooth silver surface, becomes invisible. *Middle image:* slide technique. The transparent fish Cystisoma sp. is visible in the air, which has lower refractive index than water, but invisible in water. *Lower image:* biophosphorism technique. Polyipnus laternatus fish has anti-light sources in the abdomen (*arrows*) so as, with emission of light under the appropriate angle, to provide camouflage. From Warrant & Locket [1]

Conclusions

Due to the increased need for perception of light and color in conditions of reduced lighting, aquatic organisms have developed, through evolution, different mechanisms of photoreceptor functioning than those found in humans. These mechanisms depend on the depth in which they live, since the deeper the light penetrates into the water the more it fades. In addition, due to their need for camouflage, aquatic organisms have created a variety of counterfeit mechanisms for visual perception of predators.

References

1. Warrant EJ, Locket NA (2004) Vision in the deep sea. Biol Rev 79:671–712
2. Stangos NT (2002) Κλινική Οφθαλμολογία. University Studio Press, Θεσσαλονίκη
3. Fernandez HR (1979) Visual pigments of bioluminescent and non bioluminescent deep-sea fishes. Vision Res 19:589–592
4. Johnsen S, Widder EA, Mobley CD (2001) Ultraviolet absorption in transparent zooplankton and its implications for depth distribution and visual predation. Marine Biol 138:717–730
5. De Valois KK (1977) Spatial frequency adaptation can enhance contrast sensitivity. Vision Res 17:1057–1065
6. Sowden PT, Davies IRL, Roling P (2000) Perceptual learning of the detection of features in X-ray images: a functional role for improvement in adults' visual sensitivity? J Exp Psychol Hum Percept Perform 26(1):379–390
7. Gilbert CD (1996) Learning and receptive field plasticity. Proc Natl Acad Sci U S A 93:10546–10547
8. Gislen A, Warrant EJ, Dacke M, Kroger RH (2006) Visual training improves underwater vision in children. Vision Res 46:3443–3450
9. Tachibanaki S, Shimauchi-Matsukawa Y, Arinobu D, Kawamura S (2007) Molecular mechanisms characterizing cone photoresponses. Photochem Photobiol 83(1):19–26
10. Hart NS, Lisney TJ, Marshall NJ, Collin SP (2004) Multiple cone visual pigments and the potential for trichromatic colour vision in two species of elasmobranch. J Exp Biol 207:4587–4594
11. Nilsson DE (1997) Eye design, vision and invisibility in planktonic invertebrates. In: Lenz PH, Hartline DK, Purcell JE (eds.) Zooplankton: Sensory Ecology and Physiology. Gordon & Breach Publishing Group, Newark: 149–162
12. Wagner HJ, Fröhlich E, Negishi K, Collin SP (1998) The eyes of deep sea fish. II. Functional morphology of the retina. Prog Retin Eye Res 17(4):637–685

Chapter 21
The Impact of Physicochemical and Molecular Properties in Drug Design: Navigation in the "Drug-Like" Chemical Space

Theodosia Vallianatou, Costas Giaginis, and Anna Tsantili-Kakoulidou

Abstract Physicochemical and molecular properties influence both pharmacokinetic and pharmacodynamic process, as well as drug safety, often in a conflicting way. In this aspect the current trend in drug discovery is to consider ADME (T) properties in parallel with target affinity. The concept of "drug-likeness" defines acceptable boundaries of fundamental properties formulated as simple rules of thumb, in order to aid the medicinal chemist to prioritize drug candidates. Special attention is given to lipophilicity and molecular weight, since there is a tendency for those parameters to increase in regard to complex compounds generated by new technologies, with potential consequences in bioavailability, while high lipophilicity is also associated with undesired effects. Such rules have the advantage to be very simple and are easy to interpret; however their drawback is that they do not take into consideration uncertainties in measurements and calculations as well as the receptor requirements. The case of PPARs, a nuclear receptor family, is discussed in detail in regard to the chemical space covered by the ligands, focusing on the high demands of the ligand binding domain in both lipophilicity and molecular size. Such paradigms indicate that it would be more appropriate to adapt drug-like properties according to specific drug discovery projects.

Keywords Drug discovery • Lipophilicity • Drug-likeness • PPARs

T. Vallianatou • A. Tsantili-Kakoulidou (✉)
Department of Pharmaceutical Chemistry, School of Pharmacy, University of Athens, Panepistimiopolis, Zografou, Athens 157 71, Greece
e-mail: tsantili@pharm.uoa.gr

C. Giaginis
Department of Food Science and Nutrition, University of the Aegean, 2 Mitropoliti Ioakeim Street, Myrina, Lemnos 81400, Greece

P. Vlamos, A. Alexiou (eds.), *GeNeDis 2014*, Advances in Experimental Medicine and Biology 822, DOI 10.1007/978-3-319-08927-0_21

21.1 Introduction

The impact of physicochemical properties in drug action has been rationalized since the early work of Hansch in the middle of the 1960s [1]. According to the first QSAR papers biological activity was considered as a function of lipophilicity, and electronic and steric properties [1–3]. Since those early years the evolution of the QSAR field has seen remarkable growth. Nowadays, a large arsenal of more than 4,000 molecular descriptors can be treated by sophisticated chemometric techniques to model biological activity [4]. At the same time the strategy of drug design has shifted to a more holistic approach considering absorption, distribution, metabolism, elimination (ADME) characteristics, as well as safety issues in parallel with target affinity [5, 6]. Drug candidates should possess favorable physicochemical and molecular properties that would elicit more chances for them to reach the market, reducing the attrition rate. The concept of drug-likeness associated mostly with pharmacokinetics and safety considers simple properties which should obey certain rules to guide chemical space navigation [7]. In the present paper the effect of the most important physicochemical and molecular properties is discussed in regard to both receptor requirements and pharmacokinetic process. A paradigm concerning PPAR ligands is presented.

21.2 Lipophilicity–Solubility–Ionization

The role of traditional physicochemical properties such as lipophilicity, solubility, and ionization is well established. Lipophilicity, expressed by the octanol–water partition coefficient (logP) or distribution coefficients (logD for ionizable compounds), is of paramount importance, influencing both pharmacokinetic and pharmacodynamic behavior as well as toxicological aspects [1–3, 8]. On the other hand aqueous solubility is associated with formulation and oral dosage administration, being a prerequisite for drug absorption to start through the gastrointestinal system [9]. Solubility, within the physiological pH range in conjunction with permeability, is included in the Biopharmaceutical Classification System (BCS) which categorizes drugs according to four classes [10]. BCS is of great interest for pharmaceutical industry since for drugs belonging to class I—high solubility/high permeability—bioequivalence studies are waived.

The majority of drugs contain ionizable groups, mostly basic functions. Acidic function is in certain cases an essential receptor requirement; strong acidity however is an unfavorable characteristic for permeability. Ampholytes constitute a particular class of drugs with their own physicochemical characteristics influencing their pharmacokinetic behavior in a more complex way, especially when they exist in zwitterionic form [11]. Within a series of chemically congeneric compounds however, as it is often the case in medicinal chemistry, there is no much variation in the pKa values. More to the point distribution coefficients, logD, incorporate both logP and pKa.

A huge amount of research has been conducted in regard to lipophilicity. Experimental octanol–water partition coefficients (logP) have been compiled in commercially available databases [12, 13] while several algorithms have been developed and implemented in various software for logP or logD predictions [14]. Binding to proteins correlates linearly with logP, while permeability may follow parabolic or bilinear relationships with logP or logD. Thus optimum logP values (logPo) have been defined for the penetration through certain biological barriers [15]. There is however strong evidence that high lipophilicity is associated with undesired drug features, like extensive and unpredictable metabolism, high plasma protein binding, or accumulation to tissues. The minimum hydrophobicity concept formulated in 1987 by Hansch et al. dictates that drug design should be oriented to molecules with no more lipophilicity than that required for their biological action (permeability and affinity) [15]. This principle is applicable either for compounds designed to act in the periphery as well as for molecules designed to stimulate receptors in the CNS. For the latter case there is evidence that optimum lipophilicity for BBB penetration (logP ~ 2) is related to undesirable sedative activity. More to the point hydrophobic binding by displacing water molecules from the binding pocket is nonspecific interaction increasing the chance to off-target binding and nonspecific toxicity. Ten years later upper limits for logP values [16] were proposed by Lipinski et al. who included lipophilicity in the well-known rule of 5 (RoF).

21.3 Descriptors, Descriptors, Descriptors... Which Molecular Properties Are We Looking for in Medicinal Chemistry?

Molecular descriptors are derived from different theories, such as quantum chemistry, organic chemistry, physical chemistry, computational chemistry, information theory, graph theory, and so on [4]. In addition constitutional descriptors count the presence of certain structural characteristics like heteroatoms, rings, bonds types, hydrogen bond acceptor, or donor sites. To extract relevant information from a large descriptor pool, and transform this information to knowledge incorporated in a model, different chemometric techniques are applied. The interpretation of the models however as well as the back conversion of the ensemble of the included molecular descriptors to chemical structure is not always an easy task. In fact, molecular descriptors may be further categorized to those useful for predictions of the biological activity of new compounds or for screening compound libraries and those which are easily interpretable giving insight on the mechanism of action. The latter are more familiar to the medicinal chemist and are helpful to guide further synthesis within a given chemical category. In principle, receptor–ligand interactions involve electrostatic interactions, hydrogen bonding, van der Waals and hydrophobic interactions. Further to the traditional electronic substituent constants, energy parameters like E_{HOMO}, E_{LUMO}, maximum or minimum electrostatic potential, dipole moments, partial charges, protonation states, or acidic/basic ionization constants may express the influence of electronic properties. The count of hydrogen bond acceptor and donor sites, hydrogen bond acidity and basicity parameters, and polar surface area (PSA) are among the most common descriptors to describe hydrogen bonding. Octanol–water logP as well as nonpolar surface area or volume reflects hydrophobic interactions, while bulk parameters like molecular weight, or molecular volume, polarizability, and molecular refractivity may reflect size requirements or/and (in the case of the last two) the contribution of van der Waals interactions. Flexibility usually expressed by the number of rotatable bonds is also an important straightforward interpretable property influencing binding affinity. Constitutional descriptors can describe additional structural requirements for the stabilization of drug–receptor interactions. Among these properties lipophilicity, molecular size, hydrogen bonding, PSA, and flexibility are important also in regard to bioavailability and safety issues of drug candidates, often in conflict with receptor requirements [17].

21.4 Drug-Likeness: The Development of Metrics

The evolution in synthetic possibilities and biological testing elicits new experimental data to be generated on a regular basis, following research into new chemistries and the advantages of high-throughput screening and combinatorial chemistry [18]. Such conditions and the society requirements for the development of new safer drugs faster brought forth the establishment of simple rules or metrics to guide the initial design of drug candidates from a vastly expanding chemical space. The first and still mostly applicable rule was formulated by Lipinski et al. in 1997, known as the rule of 5 (RoF) which sets upper limits (multiples of 5) for four fundamental molecular descriptors. According to RoF, molecular weight (MW) should not exceed 500 Da, calculated lipophilicity (clogP) should not exceed 5, hydrogen bond donor sites (HD) should not be more than 5, and hydrogen bond acceptor sites not more than 10. Upon pairwise violation of these limits, bioavailability problems may occur in the case of orally administered drugs [16]. RoF was further extended including cutoff values or ranges for additional properties, the most common being: PSA (<140), number of rotatable bonds (<10), Molar Refractivity (40–130), number of aromatic rings (<3), total number of atoms (20–70) [19]. Concerning safety, increased relative risk (6:1) for an adverse event in toxicology studies may be anticipated for compounds possessing both high lipophilicity (ClogP >3) and low topological polar surface area (TPSA <75 A) [20]. Stricter cutoff values are suggested for compounds active in the Central Nervous System (CNS-likeness) [21]. The development of Fragment-Based Drug

Design (FBDD) led to the establishment of the rule of 3 for lead compounds according to which $MW < 300$, $\log P < 3$, $HD < 3$, and $HA < 6$ [22].

In fact, the establishment of these rules of thumb intended to hamper the increase in size and complexity of drug candidates which were generated by the new techniques mainly the combinatorial chemistry and FBDD. The advantages of smaller and less lipophilic drug candidates were further recognized in terms of receptor binding. A new category of metrics emerged according to which affinity is normalized against molecular size, expressed as the number of heavy atoms N_H, and/or lipophilicity. Ligand Efficiency (LE) [23] and Ligand Lipophilicity Efficiency (LLE) [24] may be used to prioritize drug candidates with quasi equal potency.

$$\mathrm{LE} = \Delta G/N_H = RT \times pIC_{50}/N_H = 1.4 \times pIC_{50}/N_H$$
$$\mathrm{LLE} = pIC_{50} - \log P$$

In terms of thermodynamics, according to LLE drug–receptor interaction should be optimized in regard rather to the enthalpic component through specific interactions. Increase in entropy through hydrophobic binding increases the risk of undesired effects (next to poor ADME properties) in agreement with the minimum hydrophobicity concept.

Other metrics include: Percent Efficiency Index, PEI = f(Inh)/Mw and Binding Efficiency Index BEI = pIC_{50}/Mw normalizing activity against molecular weight expressed in kDa. The Surface Efficiency Index, SEI = pIC_{50}/PSA considers Polar Surface Area (in 100 s Å) while Ligand Efficiency Dependent Lipophilicity, LEDL = logP/LE considers both lipophilicity and size to normalize potency [23, 25].

The advantages of such metrics are that they are simple and easy to interpret. These same features however may be considered as drawbacks since such metrics can be easily "over-interpreted." In addition, they do not account for uncertainties mainly in the calculation of lipophilicity. More to the point, although they are considered as universal, they are not since they do not account for different therapeutic categories and thereupon for receptor requirements.

21.5 Navigation in Drug-Like Chemical Space: The PPAR Paradigm

Peroxisome–Proliferator Activated Receptors (PPARs) belong to the nuclear hormone receptor superfamily, which includes three subtypes PPAR-α, PPAR-β/δ, and PPAR-γ [26]. Each of these subtypes appears to be differentiated in a distinct tissue-specific manner, playing a pivotal role in glucose and lipid homeostasis. PPAR-α offers a target for the treatment of dyslipidemia, with fibrates being well-known approved drugs. PPAR-γ, the most investigated subtype, is a molecular target for the treatment of type 2 diabetes mellitus with rosiglitazone and pioglitazone of the thiazolidinedione family being marketed drugs [26, 27]. Recently considerable interest has been oriented in combining the beneficial effects of dual PPAR-α and PPAR-γ activation, according to the concept of multi-target approach, in the aim to achieve synergism and to circumvent side effects of PPAR-γ, such as weight gain, fluid retention, and edema [28]. The implication of PPARs, mainly of the β/δ and γ subtypes, in CNS disorders and in particular in Parkinson disease, multiple sclerosis, and CNS trauma injury is under investigation [29]. A particular feature of the Ligand Binding Domain (LBD) in PPARs is the very large Y-shaped cavity within the protein. It includes a very flexible entrance allowing large ligands to enter the cavity and then branches into two arms, approximately equally long [27]. Arm I is the only region with substantially polar residues, which form part of a hydrogen-bond network with the ligands upon binding. It includes the AF-2 helix that contains the transcriptional activation domain. In PPAR-δ isoform the area next to AF2 is smaller than in the other two subtypes. The hydrophobic arm II and the interior of the entrance are mainly hydrophobic. This receptor topology implies particular requirements which are reflected in the molecular properties of PPAR agonists and which may contradict the upper limits defined by the different metrics.

Tracking the property space for a large data set of PPAR-γ agonists ($n = 1{,}152$ compounds) 24 and 59 % of compounds showed violations of RoF concerning excess MW and clogP, respectively [29]. 21 % of compounds exceed both clogP and MW upper limits. Classifying the compounds according to their activity, in the case of highly active ligands ($pEC_{50} > 7$) 40 % of them exceeded both MW and clogP upper limits showing a twofold violation of RoF. Although hydrogen bonding is essential to stabilize the ligand–receptor complex no violations in the number of hydrogen bond donor and acceptor sites were observed. An analogous study restricted to two chemical categories, mainly to tyrosine and thiazolidinedione analogues, has shown however that high activity can be achieved with moderate lipophilicity [30, 31]. In contrast, in the case of indole analogues with dual PPAR-α/γ activity high demands in both lipophilicity and molecular weight are required [32]. Further comparative studies on PPAR-α/γ ligands demonstrated that high lipophilicity is more important for PPAR-α subtype, while high molecular size is imperative for PPAR-γ. Careful inspection of the differences in amino acid residues in the LBD of PPAR-α and PPAR-γ and molecular simulation supported the different impact of the two properties in binding [33]. Since however, for safety reasons, a careful balance in PPAR-α and PPAR-γ activity is desirable, drug candidates may not be selected among the most active compounds. Summarizing, the high demands on lipophilicity and molecular size of PPAR-α/γ although contradictory to drug-like properties are essential receptor requirements and should be taken into account in a consensus manner for the design of new candidates.

Conclusions

The term "drug-likeness" has gained prevalence in the drug discovery community. It defines acceptable boundaries of fundamental properties formulated as simple rules of thumb. Such rules however do not take into consideration receptor requirements, as shown in the PPAR paradigm. In this aspect it would be more appropriate to adapt drug-like properties according to specific drug discovery projects.

References

1. Hansch C, Fujita T (1964) ρ-σ-π analysis. A method for the correlation of biological activity and chemical structure. J Am Chem Soc 86:1616–1626
2. Hansch C (1969) A quantitative approach to biological structure-activity relationships. Acc Chem Res 2:232–239
3. Hansch C, Leo A (eds) (1995) Exploring QSAR: fundamentals and applications in chemistry and biology. American Chemical Society, Washington, DC
4. Todeschini R, Consonni V (2009) Molecular descriptors for chemoinformatics. In: Mannhold R, Kubinyi H, Folkers G (eds) Methods and principles in medicinal chemistry. Wiley, Weinheim
5. van de Waterbeemd H, Smith DA, Beaumont K, Walker DK (2001) Property-based design: optimization of drug absorption and pharmacokinetics. J Med Chem 244:1313–1333
6. Gaviraghi G, Barnaby RJ, Pellegatti M (2001) Pharmacokinetic challenges in lead optimization. In: Testa B, van de Waterbeemd H, Folkers G, Guy R (eds) Pharmacokinetic optimization in drug research. Verlag Helvetica Chimica Acta: Zürich and Wiley – VCH, Weinheim
7. Yusof Y, Segall MD (2013) Considering the impact drug-like properties have on the chance of success. Drug Disc Today 18:659–666
8. van de Waterbeemd H, Testa B (1987) The parametrization of lipophilicity and other structural properties in drug design. In: Testa B (ed) Advances in drug research, vol 16. Academic, New York
9. Augustijns P, Wuyts B, Hens B, Annaert P, Butler J, Brouwers J (2013) A review of drug solubility in human intestinal fluids: implications for the prediction of oral absorption. Eur J Pharm Sci. doi: 10.1016/j.ejps.2013.08.027 [Epub ahead of print]
10. Yu LX, Amidon GL, Polli JE, Zhao H, Mehta MU, Conner DP, Shah VP, Lesko LJ, Chen ML, Lee VH, Hussain AS (2002) Biopharmaceutics classification system: the scientific basis for biowaiver extensions. Pharm Res 19:921–925
11. Pagliara A, Carrupt PA, Caron G, Gaillard P, Testa B (1997) Lipophilicity profiles of ampholytes. Chem Rev 97:3385–3400
12. Howard PH, Meylan W (2000) PHYSPROP database. Syracuse Research Corp, Syracuse
13. Physicochemical Parameter Database, Medicinal Chemistry Project, Pomona College, Claremont, CA, USA. www.biobyte.com/bb/prod/cqsar.html
14. Mannhold R, Poda GI, Ostermann C, Tetko IV (2009) Calculation of molecular lipophilicity: state-of-the-art and comparison of logP methods on more than 96,000 compounds. J Pharm Sci 98:861–893
15. Hansch C, Bjorkroth JP, Leo A (1987) Hydrophobicity and central nervous system agents: on the principle of minimal hydrophobicity in drug design. J Pharm Sci 76:663–687
16. Lipinski CA, Lombardo F, Dominy BW, Feeney PJ (1997) Experimental and computational approaches to estimate solubility and permeability in drug discovery and development settings. Adv Drug Deliv Rev 23:3–25

17. Oprea T (2002) Current trends in lead discovery: are we looking for the appropriate properties? J Comput Aided Mol Des 16:325–334
18. Hertzberg RP, Pope AJ (2000) High-throughput screening: new technology for the 21st century. Curr Opin Chem Biol 4:445–451
19. Veber DF, Johnson SR, Cheng HY, Smith BR, Ward KW, Kopple KD (2002) Molecular properties that influence the oral bioavailability of drug candidates. J Med Chem 45:2615–2623
20. Hughes JD, Blagg J, Price DA, Bailey S, Decrescenzo GA, Devraj RV, Ellsworth E, Fobian YM, Gibbs ME, Gilles RW, Greene N, Huang E, Krieger-Burke T, Loesel J, Wager T, Whiteley L, Zhang Y (2008) Physiochemical drug properties associated with in vivo toxicological outcomes. Bioorg Med Chem Lett 18:4872–4875
21. Pajouhesh H, Lenz GR (2005) Medicinal chemical properties of successful central nervous system drugs. NeuroRx 2:541–553
22. Congreve M, Carr R, Murray C, Jhoti H (2003) A "Rule of Three" for fragment-based lead discovery. Drug Discov Today 8:876–877
23. Abad-Zapatero C (2007) Ligand efficiency indices for effective drug discovery. Expert Opin Drug Discov 2:469–488
24. Leeson PD, Springthorpe B (2007) The influence of drug-like concepts on decision-making in medicinal chemistry. Nat Rev Drug Discov 6:881–890
25. Keserü GM, Makara GM (2009) The influence of lead discovery strategies on the properties of drug candidates. Nat Rev Drug Discov 8:203–212
26. Willson TM, Brown PJ, Sternbach DD, Henke BR (2000) The PPARs: from orphan receptors to drug discovery. J Med Chem 43:527–550
27. Giaginis C, Theocharis S, Tsantili-Kakoulidou A (2009) Structural basis for the design of PPAR-gamma ligands: a survey on quantitative structure- activity relationships. Mini Rev Med Chem 9:1075–1083
28. Ebdrup S, Pettersson I, Rasmussen HB, Deussen H, Jensen AF, Mortensen SB, Fleckner J, Pridal L, Nygaard L, Sauerberg P (2003) Synthesis and biological and structural characterzation of the dual-acting peroxisome proliferator-activated receptor α/γ agonist ragaglitazar. J Med Chem 46:1306–1317
29. Giaginis C, Zira A, Theocharis S, Tsantili-Kakoulidou A (2008) Property distribution in the chemical space of PPAR-γ agonists: Evaluation of drug-like characteristics. Rev Clin Pharmacol Pharmacokin Intern Ed 22:366–368
30. Giaginis C, Theocharis S, Tsantili-Kakoulidou A (2007) A consideration of PPAR-γ ligands in respect to lipophilicity: current trends and perspectives. Expert Opin Investig Drugs 16:413–417
31. Giaginis C, Theocharis S, Tsantili-Kakoulidou A (2008) Application of multivariate data analysis for modeling receptor binding and gene transactivation of tyrosine-based PPAR-γ ligands. Chem Biol Drug Des 72:257–264
32. Giaginis C, Theocharis S, Tsantili-Kakoulidou A (2009) A QSAR study on indole-based PPAR-γ agonists with respect to receptor binding and transactivation data. QSAR Comb Sci 28:802–805
33. Vallianatou T, Lambrinidis G, Giaginis C, Mikros E, Tsantili- Kakoulidou A (2013) Analysis of PPAR-α/γ activity by combining 2-D QSAR and molecular simulation. Mol Inf 32:431–445

Chapter 22
Advanced Drug Delivery Nanosystems: Perspectives and Regulatory Issues

Costas Demetzos

Abstract This chapter deals with the classification of **D**rug **D**elivery **n**ano **S**ystems (**DDnSs**) with a **M**odulatory **C**ontrolled **R**elease profile (**MCR**) denoted as **M**odulatory **C**ontrolled **R**elease nano **S**ystems (**MCRnSs**). Conventional (**c**) and advanced (**a**) **DDnSs** are denoted by the acronyms **cDDnSs** and **aDDnSs**, and can be composed of a single or more than one biomaterials, respectively. The classification was based on their characteristics such as surface functionality (*f*), the nature of biomaterials used, and the kind of interactions between biomaterials. The **aDDnSs** can be classified as **Hy**bridic (**Hy**-) or **Chi**meric (**Chi**-) based on the nature—same or different, respectively—of biomaterials and inorganic materials used. The nature of the elements used for producing advanced biomaterials is of great importance and medicinal chemistry contributes effectively to the production of **aDDnSs**.

Keywords Advanced Drug Delivery nano Systems (aDDnSs) • Nanotechnology • Controlled release

Nanotechnology is the *art* of producing little devices and machines at the mesoscopic and molecular scale. Pharmaceutical Nanotechnology and Nanomedicine have branched out in different directions, each of them embodying the key insight that the ability to structure materials, devices, and systems at the molecular scale can bring enormous immediate benefits in the research of drug delivery systems and diagnostic methods. A drug delivery system (DDS) is defined as a technological formulation or a device that enables the introduction of a therapeutic substance in the body and improves its efficacy and safety by controlling the rate, time, and place of release of drugs in the body [1].

The major advances of such systems are the possibility of controlling size, structure, and morphology of the nanoassembly, as well as the possibility of integrating multifunctionalities in a single nanodevice or complex nanosystem and therefore system colloidal properties in different dispersion media, with the

C. Demetzos (✉)
Department of Pharmaceutical Technology, Faculty of Pharmacy, University of Athens, Panepistimioupolis Zografou, Athens 15771, Greece
e-mail: demetzos@pharm.uoa.gr

P. Vlamos, A. Alexiou (eds.), *GeNeDis 2014*, Advances in Experimental Medicine and Biology 822, DOI 10.1007/978-3-319-08927-0_22

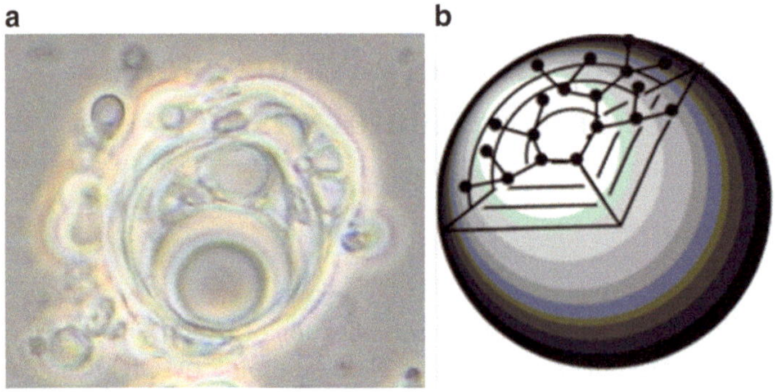

Fig. 22.1 Examples of advanced Drug Delivery nano Systems (aDDnSs). (**a**) Liposomes in Liposomes (LiLs) and (**b**) Liposomal "locked-in" dendrimers, chi-aDDnSs

aid of a large variety of physicochemical parameters, such as chemical composition, concentration, and temperature.

A **M**odulatory **L**iposomal **C**ontrolled **R**elease **S**ystem (**MLCRS**) combined liposomal and dendrimeric technologies (Fig. 22.1) incorporating the anticancer drug doxorubicin has been recently reported while **D**endrimers *locked-into* **L**iposomal formulation (**D** *locked-in* **L**) is a new term in the drug delivery and considered as a class of **MLCRSs**. We followed the concept of using acronyms, in order to reach a specific description of **MCRnSs**.

Recently our group classified the MLCRS nanocarriers as *hybridic* and *chimeric* and they are characterized as advanced Drug Delivery nano Systems (aDDnSs). Preclinical studies have shown that aDDnSs can interfere with the pharmacokinetic profile of the incorporated bioactive molecule and, consequently, alter its **ADME** (**A**bsorption, bio**D**istribution, **M**etabolism, and **E**xcretion) profile. aDDnSs can be characterized as mixed nanosystems due to the combination of different in nature bionanomaterials and are used for biomimetic delivery [2–6]. The interest in such systems stems from the possibilities for basic understanding of biological behavioral motifs, since biological systems extensively use mixed materials to create multifunctional self-assembled nanostructures, too.

In our point of view and according to the significance for producing new drug delivery systems with a controlled release profile, it is of importance to propose a classification system which can facilitate their description and harmonize the categorization of **M**odulatory **C**ontrolled **R**elease drug delivery **n**ano **S**ystems (**MCRnSs**) (Scheme 22.1.).

Drug delivery has matured from its original goal of prolonging the duration of drug release; it now involves customized systems that are designed to achieve specific spatial and temporal control; future generations of drug delivery systems will incorporate "smart" biosensing functionalities that will enable unaided in vivo feedback control. MCRnSs are considered as drug delivery systems that can modulate the release of the encapsulated drug. Conventional (c) or advanced (a) DDnSs with a modulatory controlled drug release profile are those composed

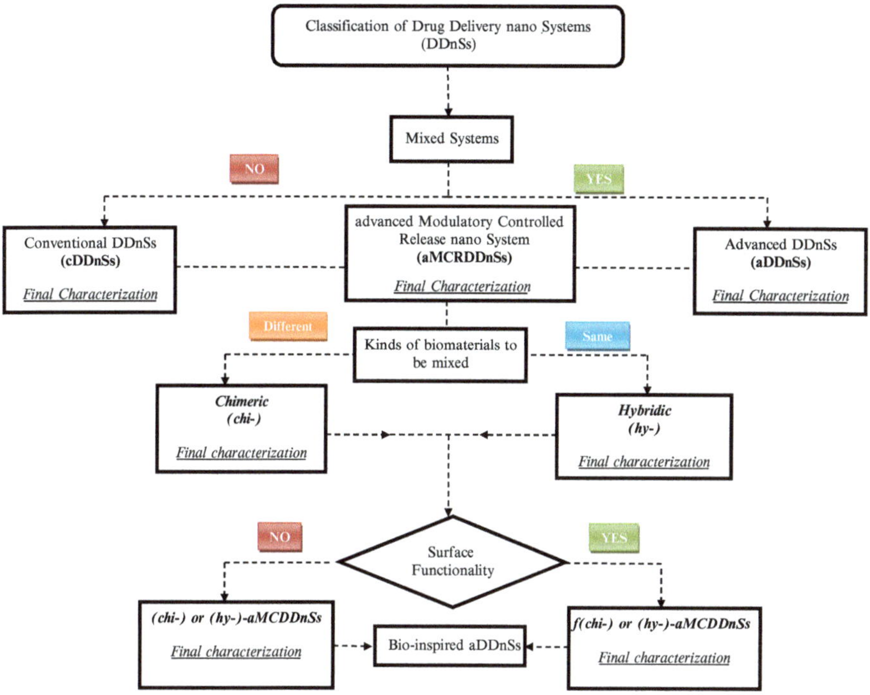

Scheme 22.1 The classification of drug delivery nanosystems

of a simple or mixed, i.e., Hybridic (Hy-) or Chimeric (Chi-), respectively bionanomaterials that can affect the release profile of the encapsulated drug, can alter the pharmacokinetic parameters, and finally improve its effectiveness. The benefits of DDnSs are related to the properties of biomaterials used which could be natural like chitosan, dextran, lipids, and phospholipids from natural sources etc, or synthetic like dendrimers, block co-polymers, lipids, etc, and to the biophysical profile of the final nanosystem. The interactions between biomaterials and the mixing process are of great importance and should be taken into consideration [7]. The production process of an aDDnSs in our point of view should be based on the kinds of biomaterials used and their properties in vivo. This could be achieved by understanding the properties of matter at nanoscale level and by easily recognizing the qualitative and quantitative properties of the drug carrier in terms of the physics evolution. The nature of the elements used for producing advanced biomaterials is of great importance and medicinal chemistry and new synthetic bioelements (i.e., new polymeric and dendritic structures) play a crucial role in the production of effective advanced Drug Delivery nano Systems (aDDnSs). However a classification in the field of drug delivery is needed and we believe that the proposal of our classification system can be used as a tool for further improvements in this subject, in order to reach a universal nomenclature of DDnSs with an MCR profile.

References

1. Gaspar R, Aksu B, Cuine A, Danhof M, Takac MJ, Linden HH, Link A, Muchitsch EM, Rowland M, Noe CR, Smith DA, Tucker GT, Crommelin DJ, Peck CC, Rocci ML Jr, Basançon L, Shah VP (2012) Impact of the pharmaceutical sciences on health care: a reflection over the past 50 years. J Pharm Sci 101:4075
2. Gardikis K, Hatziantoniou S, Bucos M, Fessas D, Signorelli M, Felekis T, Zervou M, Screttas C, Steele B, Ionov M, Scretta M, Klajnert B, Bryszewska M, Demetzos C (2010) New drug delivery nanosystem combining liposomal and dendrimeric technology (liposomal locked-in dendrimers) for cancer therapy. J Pharm Sci 99:3561
3. Pippa N, Pispas S, Demetzos C (2012) The fractal hologram and elucidation of the structure of liposomal carriers in aqueous and biological media. Int J Pharm 430:65
4. Pippa N, Pispas S, Demetzos C (2012) The delineation of the morphology of charged liposomal vectors via a fractal analysis in aqueous and biological media: physicochemical and self-assembly studies. Int J Pharm 437:264
5. Pippa N, Kaditi E, Pispas S, Demetzos C (2013) PEO-b-PCL:DPPC chimeric nanocarriers: self-assembly aspects in aqueous and biological media and drug incorporation. Soft Matter 9:4073
6. Pippa N, Merkouraki M, Pispas S, Demetzos C (2013) DPPC:MPOx chimeric advanced Drug Delivery nano Systems (chi-aDDnSs): physicochemical and structural characterization, stability and drug release studies. Int J Pharm 450:1
7. Demetzos C, Pippa N (2014) Advanced drug delivery nano systems (aDDnSs): a mini-review. Drug Deliv 21(4):250–257

Chapter 23
Bio-inspired Chimeric Drug Delivery nano Systems (Chi-DDnSs): Their Fractal Hologram and Regulatory Aspects

Natassa Pippa, Stergios Pispas, and Costas Demetzos

Pharmaceutical Nanotechnology can provide challenges for producing innovative drugs based on bio-inspired nanostructures (advanced Drug Delivery nano Systems—aDDnSs) [1]. These systems could be correlated to the living organisms due to their self-assembly properties, their hierarchical structural organization, as well as to their biocompatibility and biodegradability characteristics. aDDnSs can be considered as new therapeutic outcomes in nanomedicine that may be able to deliver pharmacomolecules to specific tissues and can improve their PK/PD (pharmacokinetics/pharmacodynamics) behavior and affect their total bioavailability. Drug delivery is a scientific sector that promotes the innovation by designing formulation approaches that can produce aDDnSs. Drug nanocarriers produced by the combination of bionanomaterials, such as liposomes, (co)polymers and dendrimers, are considered as new classes of complex self-assembled soft nanostructures. The major advantage of such bio-inspired delivery systems lies in the ability to control the size, the structure, and the morphology of the nano-assembly, due to the cooperativity of the components. The morphology of

N. Pippa
Department of Pharmaceutical Technology, Faculty of Pharmacy, National and Kapodistrian University of Athens, Panepistimioupolis Zografou, Athens 15771, Greece

National Hellenic Research Foundation, Theoretical and Physical Chemistry Institute, 48 Vassileos Constantinou Avenue, Athens 11635, Greece
e-mail: natpippa@pharm.uoa.gr

S. Pispas
National Hellenic Research Foundation, Theoretical and Physical Chemistry Institute, 48 Vassileos Constantinou Avenue, Athens 11635, Greece
e-mail: pispas@eie.gr

C. Demetzos (✉)
Department of Pharmaceutical Technology, Faculty of Pharmacy, National and Kapodistrian University of Athens, Panepistimioupolis Zografou, Athens 15771, Greece
e-mail: demetzos@pharm.uoa.gr

P. Vlamos, A. Alexiou (eds.), *GeNeDis 2014*, Advances in Experimental Medicine and Biology 822, DOI 10.1007/978-3-319-08927-0_23

nanoparticles has been recently described via fractal analysis [2–7]. The fractal dimension could be a useful tool for the development and characterization of innovative nanocarriers for pharmacomolecules with complete knowledge of their structural characteristics. This approach can disclose the pharmacokinetic reality of behavior of the candidate drugs, and can improve their therapeutic value by highlighting the interdependence of particle morphology and biological and pharmacokinetic processes. In conclusion, there is growing effort to refine methodologies for preparation and physicochemical characterization, in order for these new technologies to have the potential to meet regulatory requirements.

References

1. Demetzos C, Pippa N (2014) Advanced drug delivery nano systems (aDDnSs): a mini-review. Drug Deliv 21(4):250–257
2. Pippa N, Kaditi E, Pispas S, Demetzos C (2013) PEO-b-PCL/DPPC chimeric nanocarriers: self-assembly aspects in aqueous and biological media and drug incorporation. Soft Matter 9:4073
3. Pippa N, Merkouraki M, Pispas S, Demetzos C (2013) DPPC:MPOx chimeric advanced Drug Delivery nanosystems (chi-aDDnSs): physicochemical and structural characterization, stability and drug release studies. Int J Pharm 450(1–2):1
4. Pippa N, Kaditi E, Pispas S, Demetzos C (2013) DPPC/poly(2-methyl-2-oxazoline)-grad-poly (2-phenyl-2-oxazoline) chimeric nanostructures as potential drug nanocarriers. J Nanopart Res 15:1685
5. Pippa N, Psarommati F, Pispas S, Demetzos C (2013) The shape/morphology balance: a study of stealth liposomes via fractal analysis and drug encapsulation. Pharm Res 30(9):2385
6. Pippa N, Dokoumetzidis A, Demetzos C, Macheras P (2013) On the ubiquitous presence of fractals and fractal concepts in pharmaceutical sciences: a review. Int J Pharm 456:340
7. Pippa N, Gardikis K, Pispas S, Demetzos C (2014) The physicochemical/thermodynamic balance of advanced drug liposomal delivery systems. J Therm Anal Calorim 116(1):99–105

Chapter 24
Amphiphilic Cyclodextrin Nanoparticles for Effective and Safe Delivery of Anticancer Drugs

Erem Bilensoy

Keywords Nanoparticle • Cyclodextrin • Cancer • Tumor targeting • Paclitaxel • Camptothecin • Tamoxifen • Cell culture • Tumor-induced animal model

Cyclodextrins, enzymatic degradation products of starch, emerge as promising biomaterials for the targeted delivery of anticancer drugs to tumor tissues. Once these molecules are chemically modified with aliphatic chains to render an amphiphilic property, they are able to form nanoparticles spontaneously without the help of surfactants.

Amphiphilic cyclodextrin nanoparticles are found to be safe, non-hemolytic carriers for anticancer drugs such as tamoxifen, paclitaxel, and camptothecin which are associated with bioavailability and toxicity problems. Amphiphilic CD nanoparticles are capable of high drug encapsulation, controlled release, and stability improvement of drugs that are labile under physiological conditions.

Recently, we have also worked on novel polycationic cyclodextrins for improved cellular interaction and effective intracellular drug delivery.

Amphiphilic cyclodextrins benefit from the EPR effect being to their size and are also actively targeted by surface modification with cationic groups, PEG chains, and folate residues to target the folate receptors overexpressed in tumor cells in the body. Therefore cyclodextrin derivatives with intrinsic nanoparticle forming abilities can be considered as promising materials for safe and effective delivery of anticancer molecules of different properties to target tissues.

E. Bilensoy (✉)
Faculty of Pharmacy, Hacettepe University, Ankara, Turkey
e-mail: eremino@hacettepe.edu.tr

P. Vlamos, A. Alexiou (eds.), *GeNeDis 2014*, Advances in Experimental Medicine and Biology 822, DOI 10.1007/978-3-319-08927-0_24

Chapter 25
The Innovations in Science and Technology as a Demand for Bio-better Medicines in Europe

Costas Demetzos

Abstract The purpose of this review is to address the role of the scientific excellence of innovative medicines as the key element in the development process in Greece. The collected statistical information and data on the absorbability of funds for research of innovative medicines, diagnostics, and advanced drug delivery systems pointed out that the Greek scientists could take advantage of the "Horizon 2020" on the continuity of their investigation, whilst how the accumulation of knowledge at Greek universities and research foundations could be translated into industrial products with added value, safe and effective for the European consumers. In conclusion, this review also is considered to provide the potential benefits in order to adapt the signaling of the "Horizon 2020" for the development of a *bio*-better Europe based on scientific inspirations. This approach could be considered as an interplay between countries and even between the north and west located countries in the European landscape.

Keywords Pharmaceutics • Nanotechnology • Biotechnology • Greece • Europe • Horizon 2020

25.1 Introduction

This article is to address the role of the scientific excellence of innovative medicines as the key element in the development process in Greece and to promote the alliance between advanced nanotechnology in Pharmaceutics, biosciences, and regulatory authorities. The integration concept between science and society could be a crucial factor that forces the regulatory agencies to create more rational oriented regulatory outcomes that can effectively contribute to the development process in European

This article reflects only of the author's opinion.

C. Demetzos (✉)
Department of Pharmaceutical Technology, Faculty of Pharmacy, University of Athens, Panepistimioupolis Zografou, Athens 15771, Greece
e-mail: demetzos@pharm.uoa.gr

P. Vlamos, A. Alexiou (eds.), *GeNeDis 2014*, Advances in Experimental Medicine and Biology 822, DOI 10.1007/978-3-319-08927-0_25

Union in an economical point of view. This multidisciplinary approach has led the society as the main partner and as the basic regulator in the developing process in the European Union, because of the excellence in science (i.e., academia and research institute) and in the drug development (i.e., Pharmaceutical Industry) which reflects the high-quality approaches in prognosis, diagnosis, health care, and finally to be beneficial to the social reimbursement system [1, 2].

25.2 The European Vision for Health care in Twenty-First Century

The evolution in science and technology, and the delay of preventing the access of patient to the right therapies at the right time, offered a ground for refocusing the priorities in the European Union and in national level concerning the strategy that should be followed in the next decades. It is of interest to point out that development of *theragnostics* (which is a key part of personalized medicine and could be defined as the proposed process of diagnostic therapy for individual patients—to test them for possible reaction to taking a new medication and to tailor a treatment for them based on the test results) in terms of the technological evolution and ability to produce "*two in one*" diagnostic and therapeutic pharmaceutical products at nanoscale level is a challenge in the research and developing process. Such areas could be new achievements and new approaches by using drug or biomacromolecular delivery nanosystems to reach receptors related to well-known biomarkers. The vision of Public Private Partnership (PPP) should be to put together multidisciplinary scientific areas such as biosciences based on *-omics* and technology based on nanoscaled materials in a collaborative way that improves the effectiveness and the safety of new drugs [3]. To accelerate drug development process needs a strong collaboration between academia, pharmaceutical, and biotechnology industries as well as the regulatory authorities. One of the main problems in pharmacotherapy is the unknown behavior of the immunological system which is crucial in the case to build new vaccines. The correlation between immune system and the protection from virus or infections is unclear and the alliance between molecular biology, *-omics*, and delivery nanosystems can produce the right systemic approach, for the protection point of view.

25.3 Pharmaceutical Nanotechnology as the Key Element in the Alliance Between Bioscience and Technology for Innovative Therapies

Pharmaceutical nanotechnology is a multidisciplinary scientific platform that can provide advancements in therapeutics with accuracy, effectiveness, and safety in clinical practice. The bio-inspired Pharmaceutical nanotechnology especially in the area of drug delivery nanosystems such as targeted drug carriers inspired by the new biological targets can produce innovative delivery nanosystems that are characterized as bio-inspired drug delivery nano systems (*bio-inspired* DDnSs) [4]. These *codes* might be characterized by a high scientific impact that implements to the results from the advanced research. The exploration of the human genome and the advances in science and technology create new milestones that already exist and are considered as the main players to the development process in medicine. The implementation of the science to the classical codes at physics, biology, chemistry, and mathematics, are the real implementation path that the European and the International decision bodies should be kept in to account. However, based on the science and technology evolution the new *codes* which can get close the pharmaceutical technologies and mainly the Pharmaceutical nanotechnology and the bio- and pharmaco-sciences, based on the *-omics* point of view, should be shuttled on the Greek ancient basement which is referred as *ortho-logical* (from the Greek words *ορθός* that means correct and *λόγος, λογική* that means rationale) approach which should be followed in order to achieve success in healthcare system and to address issues concerning the regulatory authorities, clinical guidelines, bioethics, and *-omics* and to coordinate multidisciplinary scientific and technological infrastructures. The current scientific environment in terms of pharmaco-*omics* is considered as extremely complicated. It is obvious that the alliance between science based on *-omics* and nanotechnology can create new and innovative therapeutic products that have a different production approach giving rise to new regulatory directions. Finally we have to point out that the recognition that the patients opinion as the receptors of the therapeutic results should be taken into consideration in the regulation process. It is obvious that these directions facilitate the pharmacotherapy approach disclosing new horizons which can be considered as derivable to the right site for the right patient in the right time.

25.4 In Brief: The Recent Scientific Excellence in Greece

According to the published review concerning the contribution of the Greek scientific community to the total European scientific activities based on the European project Research Potential (Specific Programme Capacities-7th Framework Programme for Research of the European Union) in sectors like Biotechnology and Genetics, Agriculture, Energy and Environment, Technologies, Informatics and Communications, Materials and Life Sciences, and Human and Sociological sciences, Greek contribution is considered as very successful [5]. We point out that this European program concerns and supports the excellence of research groups which are working in Research Institutes, Universities, or other centers of excellence in research (i.e., pharmaceutical industries etc.) in the borders [*geographical and economical periphery* (**GEP**)][1] of the European Union.

[1] This terminology is proposed by the author.

The vision of this action is to promote the integration between the European Union and GEP countries, to produce an integrated scientific environment. This project named Research Potential (Specific Programme Capacities-7th Framework Programme for Research of the European Union) is a part of the FP7 European project accounting funds of 340 million euro for the period 2007–2013. The main action of this project (i.e., 7 Capacities) was REGPOT-1 which supports the excellence of research groups, in order to integrate scientific actions running in the GEP countries of the EU. Based on the SWOT analysis which promotes **S**trengths, **W**eaknesses, **O**pportunities, and **T**hreats, a strategy was developed in order to finance and support scientific infrastructures, human resources, networking and providers, and thereby to promote the scientific results and Intellectual properties. To the best of our knowledge, the published results on the sector of Health Science showed an excellent contribution of the Greek scientific groups in the literature. The European Research Council (ERC) has also gained a central place in the Europe 2020 strategy for growth and in the Innovation Union Strategy for promoting Europe's economic recovery, global competitiveness, and social prosperity. On the other hand, in Fig. 25.1, the number of publications, the number of citations, and the normalized relevant impact indicator of publications included in the eight thematic categories in relation to publications worldwide on the same topic categories of Life Sciences for the period 2000–2010 are presented in a very comprehensive way [6].

The Greek scientists occupy an important position in the number of publications and their impact on the international scientific community, especially in biosciences

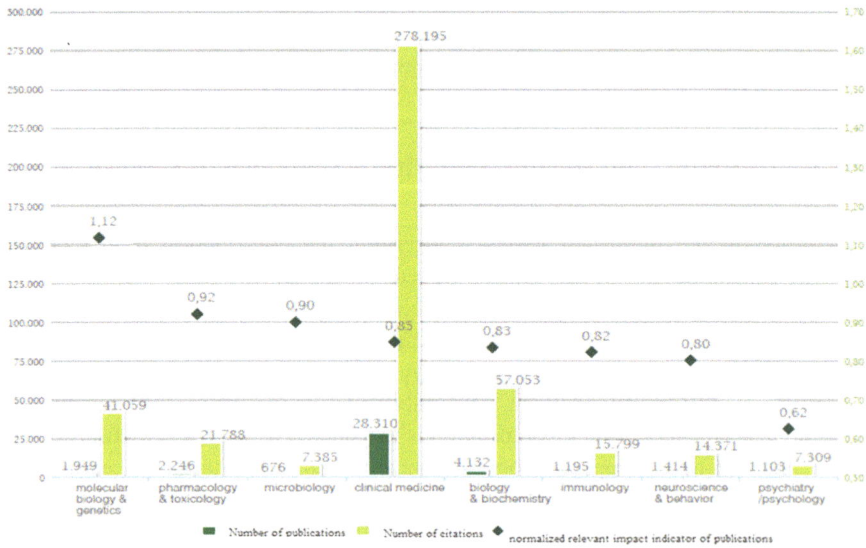

Fig. 25.1 Number of publications, number of citations, and normalized relevant impact indicator of publications included in the eight thematic categories in relation to publications worldwide on the same topic categories of Life Sciences for the period 2000–2010 (adapted from [6])

as indicated by bibliometric analysis of Greek publications in scientific journal (Fig. 25.2) [5]. According to the database of scientific publications National Science Indicators, 2010 recorded 10,219 Greek scientific publications in international journals indexes system Web of Science [7].

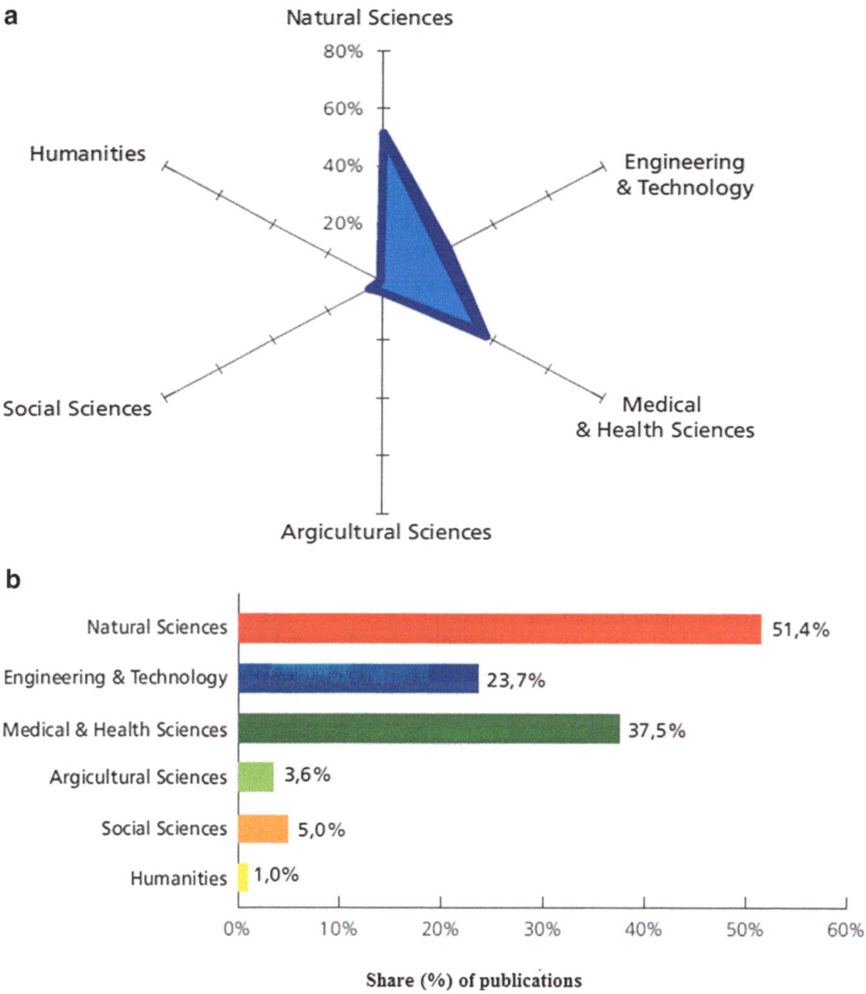

Fig. 25.2 (**a**) Breakdown of Greek publications in the six main scientific fields for the 5 years from 2004 to 2008 and (**b**) share (%) of publications (adapted from [7])

Conclusions

In conclusion, the research perspectives in Europe and the vision under Horizon 2020 are presented in this article, based on the innovation in health care and drug development. The scope of this article was to address the role of the scientific excellence and to underline that the innovations in science and technology could be the key element in the development process in Greece and in Europe as well as in the geographical and economical peripheral countries (**GEP**) in the European landscape. As the demand is the development, the science and technology mainly in health could be the driving force for the economical and social developments and for the improvements in European social reimbursement health systems. In conclusion, the purpose of the present work is to highlight the role of Greek excellence in science for developing a communication channel at European level that would flatten the economic diversification and would target only the common aspiration for producing innovative medicines with the aim of Public Health.

References

1. Gaspar R, Aksu B, Cuine A, Danhof M, Takac MJ, Linden HH, Link A, Muchitsch EM, Wilson CG, Ohrngren P, Dencker L (2012) Towards a European strategy for medicines research (2014–2020): the EUFEPS position paper on Horizon 2020. Eur J Pharm Sci 47:979
2. Crommelin DJ, Florence AT (2013) Towards more effective advanced drug delivery systems. Int J Pharm 454:496
3. Gaspar R (2007) Regulatory issues surrounding nanomedicines: settling the scene for the next generation of nanopharmaceuticals. Nanomedicine (Lond) 2:143
4. Lin Y, Mao C (2011) Bio-inspired supramolecular self-assembly towards soft nanomaterials. Front Mater Sci 5:247
5. Tzenou G, Malliou N, Sachini E (2013) Scientific excellence in EU convergence regions, 2007–2012: the case of Greece. National Documentation Centre (in Greek)
6. Sachini E, Malliou N, Houssos N (2012) Greek Scientific Publications 1996–2010: α bibliometric analysis of Greek publications in international scientific journals. National Documentation Centre
7. Sachini E, Malliou N, Houssos N, Karaiskos D (2012) Greek Scientific Publications 2000–2010 in the field of natural sciences (in Greek). National Documentation Centre

Chapter 26
The Safety of Biological Medicines for Rheumatoid Arthritis

Ioannis Papadopoulos, Costas Demetzos, Sofia Markantoni-Kyroudi, and Kyriakos Souliotis

26.1 Introduction

Biotechnological medicines have improved the treatment of various diseases and the quality of life of patients [1]. The administration of a drug may lead to adverse events that may or may not be documented. These events affect disease progression and health expenditures [2]. Rheumatoid Arthritis (RA) is an autoimmune disease for which biotechnological medicines are prescribed. RA has serious economic impact. Especially, the annual cost of RA is 41 billion euros in the USA and 45 billion euros in Europe, which are increasing dramatically with appearance of an adverse event [3].

26.2 Aims

To record the ADR of biotechnological drugs prescribed for RA., to assess their possible effects on both patient and health system, and to develop a tool aiming the efficient use of biological medicines.

I. Papadopoulos (✉) • C. Demetzos • S. Markantoni-Kyroudi
Department of Pharmaceutical Technology, Faculty of Pharmacy, University of Athens, Athens, Greece
e-mail: gianpapadopoulos@gmail.com; demetzos@pharm.uoa.gr

K. Souliotis
Faculty of Social Science, University of Peloponnese, Sparta, Greece
e-mail: soulioti@hol.gr

P. Vlamos, A. Alexiou (eds.), *GeNeDis 2014*, Advances in Experimental Medicine and Biology 822, DOI 10.1007/978-3-319-08927-0_26

26.3 Methodology

A review of existing literature on adverse events of biotechnological medicines for RA and analysis of all available data from EudraVigilance were conducted. The collection of adverse events data from rheumatology clinics of various hospitals will be conducted. Finally, any algorithm/tool currently utilized to guide the selection of the most appropriate treatment will be assessed.

26.4 Results

Sixty percent of patients are female aged 18–65 years. Twenty percent of reports indicated that an infection has occurred, 15 % highlighted the existence of inflammation at the site of injection, and 12.5 % reported a gastrointestinal disorder. The occurrence of musculoskeletal problems, hyperplasia, and nervous disorder estimated at 10 % of the reports and the respiratory problems approximately 8 %.

Conclusion
Infection or inflammation at the site of injection is the most common adverse events altering microbiological profile of patient. A possible relationships/associations between certain parameters (e.g., genome and the appearance of an adverse event) which will contribute to the safer and efficient use of biological drugs will be examined. In addition, the total average medical costs were reported to range from 5,720$ to 5,822$, while the average number of days absent from work due to a person's RA was reported to range from 2, 7 to 30 days/years. The appearance of adverse events increases dramatically the therapeutic costs.

References

1. Takeuchi T (2011) Revolutionary change in Rheumatoid arthritis management with biological therapy. Keio J Med 60:3
2. Pirmohamed M (2002) Adverse drug reactions – back to future. Br J Clin Pharmacol 55:486–492
3. Furneri G (2012) Systemic literature review on economic implications and pharmacoeconomics issues of rheumatoid arthritis. Clin Exp Rheumatol 30:73

Chapter 27
The Fractal Analysis as a Complementary Approach to Predict the Stability of Drug Delivery nano Systems (DDnSs) in Aqueous and Biological Media: A Regulatory Proposal or a Dream?

Costas Demetzos and Natassa Pippa

The morphology of drugs' nanocarriers could be considered as their *think tank,* and can be correlated with their functionality, which is mainly shuttled on their surface in which most of the interactions and interfacial phenomena are occurred. In our point of view, the *gold standard technique* for characterizing the morphology of nanoparticles could be the fractal analysis. From the experimental point of view, the determined fractal dimension illustrates in a comprehensive way the self-assembly and the morphological complexity of conventional and *chimeric* liposomal drug nanocarriers [1–5].

Additionally, fractal dimension plays an important role for the elucidation of morphological characteristics, while size and/or size distribution of drug nanocarriers did not change by changing the colloidal parameters, like temperature and concentration [6, 7]. The quantification of physical picture of the studied liposomal nanoparticles, ω_D (ω from the last letter of Greek alphabet) could be determined by the following formula:

$$\omega_D = |f - m| \tag{27.1}$$

where f is the quantification of the surface functionality via fractal dimension of nanoparticles in biological medium (i.e., fetal bovine serum, human plasma etc.) and m is the quantification of the surface morphology via fractal nanoparticles in aqueous medium (i.e., HPLC-grade water, phosphate buffer saline etc.). By using the fractal approach, the Biotechnological/Pharmaceutical Industry can disclose the effectiveness of biosimilar drugs to claim their usefulness to the patient by reducing their side effects, mainly the immunogenicity which is correlated with their stability.

C. Demetzos (✉) • N. Pippa
Department of Pharmaceutical Technology, Faculty of Pharmacy, National and Kapodistrian University of Athens, Panepistimioupolis Zografou, Athens 15771, Greece
e-mail: demetzos@pharm.uoa.gr

P. Vlamos, A. Alexiou (eds.), *GeNeDis 2014*, Advances in Experimental Medicine and Biology 822, DOI 10.1007/978-3-319-08927-0_27

References

1. Pippa N, Kaditi E, Pispas S, Demetzos C (2013) PEO-b-PCL/DPPC chimeric nanocarriers: self-assembly aspects in aqueous and biological media and drug incorporation. Soft Matter 9:4073
2. Pippa N, Merkouraki M, Pispas S, Demetzos C (2013) DPPC:MPOx chimeric advanced Drug Delivery nanosystems (chi-aDDnSs): physicochemical and structural characterization, stability and drug release studies. Int J Pharm 450(1–2):1
3. Pippa N, Kaditi E, Pispas S, Demetzos C (2013) DPPC/poly(2-methyl-2-oxazoline)-grad-poly (2-phenyl-2-oxazoline) chimeric nanostructures as potential drug nanocarriers. J Nanopart Res 15:1685
4. Pippa N, Psarommati F, Pispas S, Demetzos C (2013) The shape/morphology balance: a study of stealth liposomes via fractal analysis and drug encapsulation. Pharm Res 30(9):2385
5. Pippa N, Dokoumetzidis A, Demetzos C, Macheras P (2013) On the ubiquitous presence of fractals and fractal concepts in pharmaceutical sciences: a review. Int J Pharm 456:340
6. Pippa N, Gardikis K, Pispas S, Demetzos C (2014) The physicochemical/thermodynamic balance of advanced drug liposomal delivery systems. J Therm Anal Calorim 116:99–105
7. Demetzos C, Pippa N (2014) Advanced drug delivery nano systems (aDDnSs): a mini-review. Drug Deliv 21:250

Chapter 28
Nanothermodynamics Mediates Drug Delivery

Aikaterina L. Stefi, Evangelia Sarantopoulou, Zoe Kollia, Nikolaos Spyropoulos-Antonakakis, Athanasia Bourkoula, Panagiota S. Petrou, Sotirios Kakabakos, Georgios Soras, Panagiotis N. Trohopoulos, Alexey S. Nizamutdinov, Vadim V. Semashko, and Alkiviadis Constantinos Cefalas

Abstract The efficiency of penetration of nanodrugs through cell membranes imposes further complexity due to nanothermodynamic and entropic potentials at interfaces. Action of nanodrugs is effective after cell membrane penetration. Contrary to diffusion of water diluted common molecular drugs, nanosize imposes an increasing transport complexity at boundaries and interfaces (e.g., cell membrane). Indeed, tiny dimensional systems brought the concept of "nanothermodynamic potential," which is proportional to the number of nanoentities in a macroscopic

A.L. Stefi • E. Sarantopoulou (✉) • Z. Kollia • N. Spyropoulos-Antonakakis
National Hellenic Research Foundation, Theoretical and Physical Chemistry Institute,
48 Vassileos Constantinou Avenue, Athens 11635, Greece
e-mail: kstefi@eie.gr; esarant@eie.gr; zkollia@eie.gr; nspyropoulos@eie.gr

A. Bourkoula • P.S. Petrou • S. Kakabakos
N.C.S.R. "Demokritos", Institute for Nuclear and Radiological Sciences, Energy,
Technology and Safety, Patriarchou Grigoriou Street, Athens 15310, Greece
e-mail: nancyb@rrp.demokritos.gr; ypetrou@rrp.demokritos.gr; skakab@rrp.demokritos.gr

G. Soras
RAFARM S.A., 12 Korinthou Street, N. Psychico, 15451, Athens, Greece
e-mail: g.soras@rafarm.gr

P.N. Trohopoulos
CosmoPHOS Ltd, 77 Tsimiski Street, Thessaloniki, 54622, Greece
e-mail: panagiotis.trohopoulos@cosmophos.com

A.S. Nizamutdinov • V.V. Semashko
Kazan Federal University, 18 Kremljovskaja Street, Kazan 420008, Russia
e-mail: ua4pcy@mail.ru

A.C. Cefalas
National Hellenic Research Foundation, Theoretical and Physical Chemistry Institute,
48 Vassileos Constantinou Avenue, Athens 11635, Greece

Visiting Professor, Kazan Federal University, 18 Kremljovskaja Street,
Kazan 420008, Russia
e-mail: ccefalas@eie.gr

P. Vlamos, A. Alexiou (eds.), *GeNeDis 2014*, Advances in Experimental Medicine and Biology 822, DOI 10.1007/978-3-319-08927-0_28

system, from either the presence of surface and edge effects at the boundaries of nanoentities or the restriction of the translational and rotational degrees of freedom of molecules within them. The core element of nanothermodynamic theory is based on the assumption that the contribution of a nanosize ensemble to the free energy of a macroscopic system has its origin at the excess interaction energy between the nanostructured entities. As the size of a system is increasing, the contribution of the nanothermodynamic potential to the free energy of the system becomes negligible. Furthermore, concentration gradients at boundaries, morphological distribution of nanoentities, and restriction of the translational motion from trapping sites are the source of strong entropic potentials at the interfaces. It is evident therefore that nanothermodynamic and entropic potentials either prevent or allow enhanced concentration very close to interfaces and thus strongly modulate nanoparticle penetration within the intracellular region. In this work, it is shown that nano-sized polynuclear iron (III)-hydroxide in sucrose nanoparticles have a nonuniform concentration around the cell membrane of macrophages in vivo, compared to uniform concentration at hydrophobic prototype surfaces. The difference is attributed to the presence of entropic and nanothermodynamic potentials at interfaces.

28.1 Introduction

Nanomedicine is an innovating and a challenging scientific area of interest. It consists of the interdisciplinary combination [1] of nanotechnology and medicine. Nanomedicine includes the development of nanoparticles and techniques for diagnosis and treatment of many diseases at the nanoscale [2–5]. One of its major sectors is nanoparticle drug delivery [1, 3, 6]. Nanotechnology-based drug delivery systems have emerged as promising platforms for targeting and destroying mainly cancer cells [7–9] via photodynamic therapy (PDT) that uses a combination of light sources, photosensitizers such as phthalocyanines, and specific targeting molecules and nano-carriers for localized therapies at the micro/nanoscale. Following light activation of photosensitizers, reactive oxygen species are released and eventually destroy cancer cells [10], after successful intracellular internalization via endocytosis and subsequent accumulation in the endosome/lysosome compartments of the cell. Although that part of nanomedicine is highlighted as one of the most promising field, it faces with two crucial problems. Despite complex design of nano-carriers, such as synthesis of polyion complex micelles via electrostatic interactions between phthalocyanine complexes and various block copolymers [11], uncontrolled nanodrug aggregation and inefficient cell penetration is always present in most nanodrug delivery systems.

Almost all nanoparticles when come in touch with their target start to aggregate [12–15]. For example the aggregate size of SiO_2 nanoparticles increases when placed on colloidal solution. The aggregation of nanoparticles at interfaces incommodes their entrance through the cellular membrane and thus into the internal part of the cell that is the aim of many drug delivery treatments. Indeed, the cellular energy unit, the mitochondrion, is the target of many nanoparticles synthesized for the treatment of well-known diseases such as Alzheimer's disease, cancer, and metabolic disorders [16, 17]. Hence, the thermodynamic aggregation of nanoparticles at the boundaries of the cells could lead to inefficacy arriving on their target (e.g. organelle, metabolic procedure, etc.) despite that they were designed appropriately.

The second constrain in a nanomedicine project is via to the presence of a large amount of electric charge, as most of the nanoparticles used for drug delivery treatment are highly negatively charged. For example, ZnPc is negatively charged [18, 19] as also some categories of dendrimers that could be used as carriers [20]. At this point, it must be mentioned that the cellular membrane is negatively charged because of its structure [21]. Indeed, the head groups of bilayer phospholipids consist of two anionic lipids, phosphatidylserine (PS) and phosphatidylinositol (PI), that are negatively charged [22]. Consequently, at the cellular interface a repulsive force is generated and then the negatively charged nanoparticles do not even reach the cellular membrane.

On the other hand, Tiny dimensional systems introduce the concept of "nanothermodynamic potential" [23] which is proportional to the number of nanoentities in a macroscopic system, from either the presence of surface and edge effects at the interfaces or the restriction of the translational and rotational degrees of freedom of molecules within them that specify an entropic change. In the case of nanothermodynamic potentials, the core element of the thermodynamic potential is based on the assumption that the contribution of a nanosize ensemble to the free energy of a macroscopic system has its origin at the excess interaction energy between the nanostructured entities. As the size of a system is increasing, the contribution of the nanothermodynamic potential to the free energy of the system becomes negligible. Contrary to stronger interactions, which are independent of system's size, weak physical interactions are dependent on the local morphological features and the various constrains that might restrict the molecular translational motion at the nanoscale. However, the experimental demonstration of nanothermodynamic potentials is difficult to be implemented as sophisticated experimental approaches are required [24].

The nanothermodynamic potential is directly correlated to internal stress, which is responsible for a plurality of morphological structures and shapes at the nano/microscale and therefore it plays a major role in Natural evolution. Indeed, a sound example is a spontaneous generation of complexity in the form of self-assembled structures, the outcome of overstressing during a dynamical transformation of a system attempting to attain a state of *minimum* energy and *maximum* entropy [25].

Particularly, the stress–strain response at interfaces is specified, besides the chemical composition and the specific geometrical confrontations, by both the nature of the molecular forces between the boundaries and the difference of the internal, chemical, entropic, and surface components of the free energy of the interface system [26]. The challenging situation in the case of weak interactions within a specific system, e.g., hydrogen polar interactions, where the order of magnitude of free energy difference from physisorption in thin layers is in the *pJ* range and the associated strained field does not exceed a few Ångstroms, is the theoretical description of the energy transfer and the experimental verification of the internal energy differences. Any description should take into account both the molecular interactions and the conformational configuration of the combined specific system, as well as any additional constraint that might be important, such as the size and the concentration of nanocomposites. Finally, besides thermodynamic potential, an additional key theme is the interplay between the combinatorial entropic changes, including any possible restriction of the translational symmetry of molecules at the nanoscale.

In this work, nano-sized particles of polynuclear iron (III)-hydroxide in sucrose are deposited on macrophages' cells in vivo as also on the surface of hydrophobic silicon wafers and then the size and concentration of the created nano-aggregations on the interfaces between the two substances in the two cases are studied using Atomic Force Microscopy (AFM). The results emphasize the role of thermodynamics in intracellular nanodrug delivery.

28.2 Experimental

Here, THP-1 cell line was used. Cells of that cell line are monocyte-like and they were derived from a 1-year-old boy with leukemia (ATCC®TIB-202). Cell culture was performed in an incubator at a 95 % w/w ambient air, 5 % w/w CO_2 gaseous environment, and the temperature is maintained at 37 °C. The medium was RPMI 1640 serum with 2 mM L-glutamine adjusted to contain 1.5 g/L sodium bicarbonate, 4.5 g/L glucose, and 10 mM HEPES. Generally, the THP-1 cell line presents a quite quick rate of proliferation and it is suspended. Subsequently, the cells are deposited on silicon wafers coated with Poly-L-Lysine (PLL) to make them adherent. Next, silicon wafers were poured with 5 μL of solution polynuclear iron (III) hydroxide in sucrose (Venofer®, 20 mg Fe/mL in 30 % w/v sucrose solution at a pH of about 10.5–11.1). For the AFM measurements, wafers are left to dry and then fixed with 4 % paraformaldehyde (PFA).

An AFM (Bruker di Innova) was used to evaluate the morphology and roughness of the samples that contain polynuclear iron (III) hydroxide in sucrose diluted on macrophages. The AFM images were acquired in ambient conditions, in tapping mode by a phosphorus-(n)-doped silicon cantilever (Bruker, RTESPA) with a nominal spring constant of 40 N/m at 300 kHz resonance frequency and nominal radius of 8 nm. The AFM images were obtained at different scanning areas at a maximum scanning rate of 0.5 Hz and with an image resolution of 512 × 512 pixels. The AFM was placed on a vibration isolation table and inside an acoustic isolation chamber.

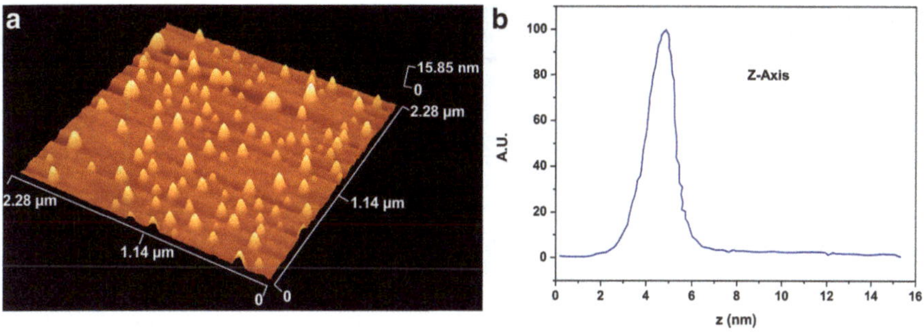

Fig. 28.1 (**a**) 3-D AFM image of agglomeration of polynuclear iron (III)-hydroxide nanoparticles in sucrose on Si wafers. (**b**) Size distribution of polynuclear iron (III)-hydroxide nanoparticles in sucrose along the z axis showing a Gaussian distribution

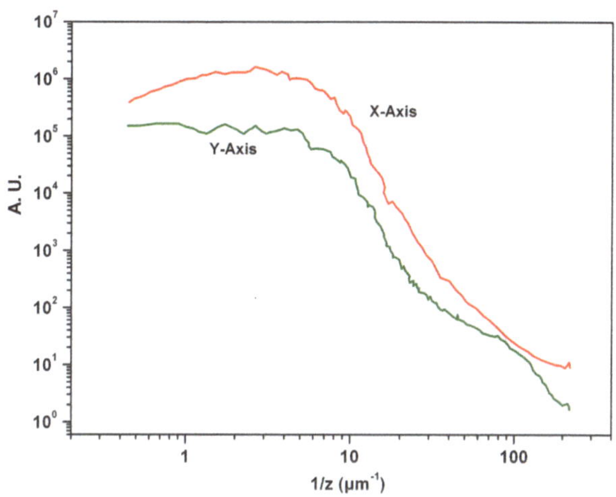

Fig. 28.2 Fast Fourier transform of polynuclear iron (III)-hydroxide particles in sucrose along the x and y axes. The Gaussian and FFT distributions indicate homogeneous thermodynamic potentials at the interface between the Si surface and the nanoparticles

28.3 Results

In the AFM image of Fig. 28.1a spherical nanoparticles of the deposited polynuclear iron (III)-hydroxide are evenly distributed on the Si surface having an average size in the $X - Y$ plane of 70 nm. The size distribution of polynuclear iron (III)-hydroxide nanoparticles in sucrose along the z axis follows the Gaussian distribution as it is shown in the histogram of Fig. 28.1b. It is obvious that there is a uniform spatial distribution of thermodynamic potential gradients between the Si surface and the nanoparticles. Additionally, analyzing with Fast Fourier transform the polynuclear iron (III)-hydroxide particles in sucrose on Si along the x and y axes homogeneous thermodynamic potentials at the interface between the Si surface and the nanoparticles were indicated via the Gaussian and FFT distributions, Fig. 28.2. On the contrary, distribution of nanoparticles among macrophage cells (THP-1 cell line) is uneven, Fig. 28.3a, suggesting the presence of thermodynamical potential at

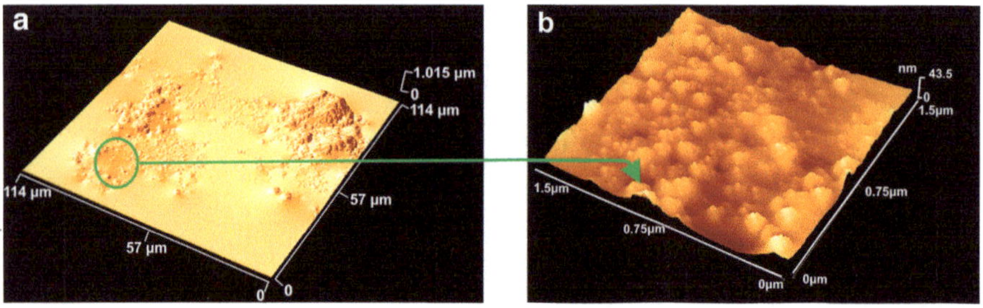

Fig. 28.3 (**a**) Accumulation of iron (III)-hydroxide nanoparticles in sucrose near the boundaries of macrophage cell in vivo that indicates the presence of nonuniform thermodynamic gradients at the cell's external wall. (**b**) A detail (1.5 × 1.5 µm) for a low roughness area of Fig. 28.3a on the macrophage cell

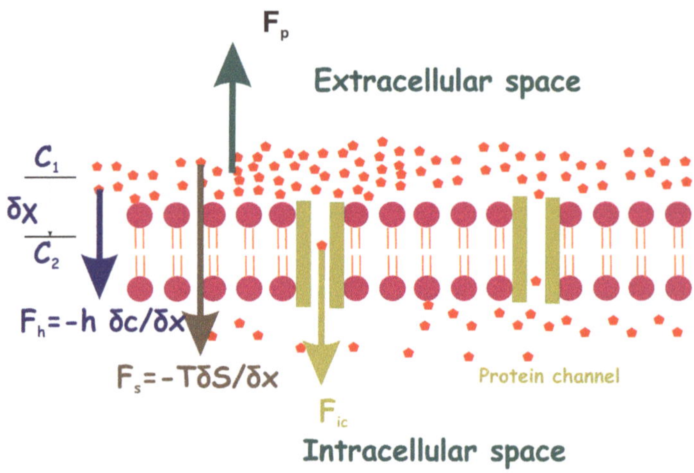

Fig. 28.4 Schematic layout of thermodynamic potential gradients near the interface of a cell. The accumulation of nanoparticles at the interface creates an entropic (F_s) and a nanothermodynamic force (F_h) across the interface. Nanoparticles are then forced inwards. A polar force (F_p) from negatively charged nanoparticles diverts them far away from the membrane

the interfaces. In Fig. 28.3b the same trend is displayed in an area of the macrophage cell even though the roughness of the aggregated nanodrugs is much lower.

Besides the chemical potential gradient near interfaces that forces particle motion from high to low concentration areas, nanoparticles brought the concept of "nanothermodynamic potential," which is proportional to the number of nanoentities in a macroscopic system or the restriction of the translational and rotational degrees of freedom of molecules, which is the source of a nanothermodynamic entropic potential. The free energy change δG at the interface between the membrane of a cell and the extra cellular medium, Fig. 28.4, under isobaric isothermal conditions is given by the relation:

$$\delta G = \delta U - T\delta(S) - h_i\delta(c_j) - \mu_i\delta(c_i) \tag{28.1}$$

where $\delta(U)$ is the internal energy change from polar interactions, $\delta(S)$ is the entropic energy change from trapping of nanoparticles on the external wall of the cell [27], h_i and c_i are the Hill's potential and the concentration of nano-sized particles, μ_i and c_i are the chemical potential and the concentration of macro-sized particles. The spatial variation of free energy $-\frac{\delta G}{\delta x}$ is the total force F exerted on the nanoparticles at the boundary of membrane and in the absence of macro-sized particles, Fig. 28.4, and is given by the equation,

$$F = -\frac{\delta G}{\delta x} = -\frac{\delta U}{\delta x} + T\frac{\delta S}{\delta x} + h\frac{\delta c}{\delta x} = F_p - F_s - F_h \tag{28.2}$$

where F_p, F_s, F_h are the electrostatic force (repulsive or attractive) between the nanoparticles and the membrane, the entropic force due to immobilization of nanoparticles at the interface, and the nanothermodynamic force due to the concentration of nanoparticles at the interface respectively, Fig. 28.4. The index j is omitted from c and h as Eq. (28.2) contains only terms which are related to nano-sized particles. The entropic and nanothermodynamic forces are directed inwards the membrane wall, while the electrostatic force of both negatively charged nanoparticles and the membrane is directed outwards the wall. Therefore both the entropic and nanothermodynamic potentials mediate nanoparticle intracellular transport within living cells.

Conclusions

Nano-sized particles of polynuclear iron (III)-hydroxide in sucrose have an increased concentration at the boundaries of the cell membrane of macrophages in vivo, compared to a uniform concentration on hydrophobic silicon wafers. The difference is attributed to the presence of entropic and nanothermodynamic potentials at interfaces that enhances aggregation of nanoparticles at interfaces.

Acknowledgments Partial financial support from the European Union, under the FP7-NMP-2012-LARGE-6 "CosmoPhos-Nano" project (reference number: 310337), and from the Russian Government under the Grand No. 02.A03.21.0002 is gratefully acknowledged.

References

1. Sanhai WR, Sakamoto JH, Canady R et al (2008) Seven challenges for nanomedicine. Nat Nanotechnol 3:242–244
2. Sousa AA, Kruhlak MJ (2013) Introduction: nanoimaging techniques in biology. Methods Mol Biol 950:1–10
3. Riehemann K, Schneider SW, Luger TA et al (2009) Nanomedicine: challenge and perspectives. Angew Chem Int Ed Engl 48(5):872–897

4. Tasciotti E, Liu X, Bhavane R et al (2008) Mesoporous silicon particles as a multistage delivery system for imaging and therapeutic applications. Nat Nanotechnol 3:151–157
5. Peer D, Karp JM, Hong S et al (2007) Nanocarriers as an emerging platform for cancer therapy. Nat Nanotechnol 2:751–760
6. LaVan DA, McGuire T, Langer R (2003) Small-scale systems for in vivo drug delivery. Nat Biotechnol 21(10):1184–1191
7. Wu Y, Sefah K, Liu H et al (2010) DNA aptamer–micelle as an efficient detection/delivery vehicle toward cancer cells. P Natl Acad Sci U S A 107:5–10
8. Dhar S, Gu FX, Langer R et al (2008) Targeted delivery of cisplatin to prostate cancer cells by aptamer functionalized Pt(IV) prodrug-PLGA-PEG nanoparticles. P Natl Acad Sci U S A 105:17356–17361
9. Gu F, Zhang L, Teply BA et al (2008) Precise engineering of targeted nanoparticles by using self-assembled biointegrated block copolymers. P Natl Acad Sci U S A 105:2586–2591
10. Dolmans DE, Fukumura D, Jain RK et al (2003) Photodynamic therapy for cancer. Nat Rev Cancer 3:380–387
11. Jang WD, Nakagishi Y, Nishiyama N et al (2006) Polyion complex micelles for photodynamic therapy: incorporation of dendritic photosensitizer excitable at long wavelength relevant to improved tissue-penetrating property. J Control Release 113:73–79
12. Yang Z, Kang S, Zhou R (2014) Nanomedicine: de novo design of nanodrugs (review article). Nanoscale 6:663–677
13. Jassby D (2011) Impact of the particle aggregation on nanoparticle reactivity, Department of Civil and Environmental Engineering. Dissertation, Duke University
14. Pranami G (2009) Understanding nanoparticle aggregation. Dissertation, Iowa State University, Ames, Paper 10859
15. Zhang XF, Xu HJ (1993) Influence of halogenation and aggregation on photosensitizing properties of zinc phthalocyanine (ZnPC). J Chem Soc Faraday Trans 89:3347–3351
16. Siddiqui MA, Alhadlaq HA, Ahmad J et al (2013) Copper oxide nanoparticles induced mitochondria mediated apoptosis in human hepatocarcinoma cells. PloS One 8(8):e69534
17. Marrache S, Dhar S (2012) Engineering of blended nanoparticle platform for delivery of mitochondria-acting therapeutics. P Natl Acad Sci U S A 109:16288–16293
18. Mutter AC, Norman JA, Tiedemann MT et al (2014) Rational design of a zinc phthalocyanine binding protein. J Struct Biol 185:178–185
19. Kažukauskas V, Arlauskas A, Pranaitis M et al (2010) Conductivity, charge carrier mobility and ageing of ZnPc/C60 solar cells. Opt Mater 32(12):1676–1680
20. Thiagarajan G, Greish K, Ghandehari H (2013) Charge affects the oral toxicity of poly (amidoamine) dendrimers. Eur J Pharm Biopharm 84(2):330–334
21. Magalhaes MAO, Glogauer M (2010) Pivotal advance: phospholipids determine net membrane surface charge resulting in differential localization of active Rac1 and Rac2. J Leukoc Biol 87(4):545–555
22. Yeung T, Gilbert GE, Shi J et al (2008) Membrane phosphatidylserine regulates surface charge and protein localization. Science 319(5860):210–213
23. Trepagnier EH, Jarzynski C, Ritort F et al (2004) Experimental test of Hatano and Sasa's nonequilibrium steady-state equality. P Natl Acad Sci U S A 101:15038–15041
24. Carberry DM, Reid JC, Wang GM et al (2004) Fluctuations and irreversibility: an experimental demonstration of a second-law-like theorem using a colloidal particle held in an optical trap. Phys Rev Lett 92:140601
25. Park BJ, Furst EM (2010) Fluid-interface templating of two-dimensional colloidal crystals. Soft Matter 6:485–488
26. Sarantopoulou E, Kollia Z, Cefalas AC et al (2008) Surface nano/micro functionalization of PMMA thin films by 157 nm irradiation for sensing applications. Appl Surf Sci 254:1710–1719
27. Cefalas AC, Sarantopoulou E, Kollia Z et al (2012) Entropic nanothermodynamic potential from molecular trapping within photon induced nano-voids in photon processed PDMS layers. Soft Matter 8:5561–5574

Chapter 29
The Neurodegenerative Potential of Chronic Stress: A Link Between Depression and Alzheimer's Disease

Ioannis Sotiropoulos

Keywords Alzheimer's disease • Chronic stress • Depression • Synaptic pathology

Modern lifestyle places individuals under increasingly greater loads of psychological and physical stress. Although the mechanisms that are triggered by stress are primarily adaptive to facilitate homeostasis, chronic stress can become maladaptive. Specifically, stress and its primary manifestation, glucocorticoid (GC) secretion, is strongly associated with neuronal atrophy/dysfunction, impaired cognition, and mood and affective disorders such as depression while a causal role of chronic stress in the etiopathology of Alzheimer's disease (AD) has been also suggested. Although cumulative evidence suggests a continuum between depres- sion and AD, and stress is suggested to play a detrimental role in both diseases, considerably less attention has been given to the suggested role of stress as a connecting risk factor. Using both transgenic and non-transgenic animals, we investigate the sequential interrelationships between these various pathogenic ele- ments, in particular with respect to the mechanisms through which stress might precipitate brain pathology. Our studies show that stress and GC trigger APP misprocessing towards the production of neurotoxic amyloid-β (Aβ) as well as abnormal tau hyperphosphorylation and aggregation resulting in synaptic malfunc- tion and atrophy that leads to the associated cognitive (memory) impairments as well as mood deficits. In addition, we demonstrated increased vulnerability of female brain to the neurodegenerative potential to stress through this ability to regulate molecular chaperones related to formation of neurotoxic Tau aggregations showing that Tau protein and

I. Sotiropoulos (✉)
Life and Health Sciences Research Institute (ICVS), School of Health Sciences, University of Minho, Braga, Portugal

ICVS/3Bs - PT Government Associate LaboratoryBraga/Guimaraes, Portugal

Department of Pharmacology, Medical School, University of AthensAthens, Greece
e-mail: sotiropoulosjohn@hotmail.com

P. Vlamos, A. Alexiou (eds.), *GeNeDis 2014*, Advances in Experimental Medicine and Biology 822, DOI 10.1007/978-3-319-08927-0_29

its malfunction has an essential role in detrimental role of chronic stress on brain and neuronal function providing molecular, electro- physiological, morphostructural, and behavioral evidence. These studies suggest an essential role of Aβ and Tau in a critical mechanism(s) through which stress and GC exert their neuro-remodelling and neurodegenerative effects upon the substrates of cognition and emotion.

Index

P. Vlamos, A. Alexiou (eds.), *GeNeDis 2014*, Advances in Experimental Medicine and Biology 822, DOI 10.1007/978-3-319-08927-0

MIX
Papier aus verantwortungsvollen Quellen
Paper from responsible sources
FSC® C105338

If you have any concerns about our products,
you can contact us on
ProductSafety@springernature.com

In case Publisher is established outside the EU,
the EU authorized representative is:
Springer Nature Customer Service Center GmbH
Europaplatz 3, 69115 Heidelberg, Germany

Printed by Libri Plureos GmbH
in Hamburg, Germany